WITHDRAWN FROM
TCC LIBRARY

TALLAHASSEE
LIBRARY
COMMUNITY COLLEGE

D0734416

WHAT
IS DEATH?

WHAT
IS DEATH?

A Scientist Looks at the Cycle of Life

Tyler Volk

A PETER N. NEVRAUMONT BOOK

John Wiley & Sons, Inc.

Copyright © 2002 by Tyler Volk. All rights reserved

Published by John Wiley & Sons, Inc., New York
Published simultaneously in Canada

No part of this publication may be reproduced, stored in a retrieval system, or transmitted in any form or by any means, electronic, mechanical, photocopying, recording, scanning, or other-wise, except as permitted under Section 107 or 108 of the 1976 United States Copyright Act, without either the prior written permission of the Publisher, or authorization through payment of the appropriate per-copy fee to the Copyright Clearance Center, 222 Rosewood Drive, Danvers, MA 01923, (978) 750-8400, fax (978) 750-4744. Requests to the Publisher for permission should be addressed to the Permissions Department, John Wiley & Sons, Inc., 605 Third Avenue, New York, NY 10158-0012, (212) 850-6011, fax (212) 850-6008, email: PERMREQ@WILEY.COM.

This publication is designed to provide accurate and authoritative information in regard to the subject matter covered. It is sold with the understanding that the publisher is not engaged in rendering professional services. If professional advice or other expert assistance is required, the services of a competent professional person should be sought.

Library of Congress Cataloging-in-Publication Data:
Volk, Tyler.
  What is death? : a scientist looks at the cycle of life / Tyler Volk.
    p. cm.
  Includes bibliographical references and index.
  ISBN 0-471-37544-6 (cloth)
  1. Death.   I. Title.

QP87.V65 2000
571.9'39—dc21

                                          00-059431

Created and Produced by
NEVRAUMONT PUBLISHING COMPANY
New York, New York

Ann J. Perrini, President
Book design: Frances White

Printed in the United States of America

10 9 8 7 6 5 4 3 2 1

To my parents,
JOSEPH AND VIVIAN,
foremost in my gratitude.

C O N T E N T S

INTRODUCTION: DEATH, THUS LIFE

We age, and most of us come to accept the persistent specter of death as an inevitable part of being alive. It's the price we disburse at the end, a price for the gift of life.

We don't normally revel in this state of affairs, of course. I, for one, wouldn't mind cheating the game. But it's futile to think about playing without paying. Though the advances of medical science often do lengthen our term of ephemerality, they cannot promise us eternity. In short, we are faced with a simple fact: "life, thus death."

This simple phrase—life, thus death—summarizes my core theme: the bond between life and death. But I am primarily intrigued with how this bond can be expressed (and perhaps far better expressed) by reversing the phrase. Let's flip the logic around and say "death, thus life."

How can death precede life? Are there cases in which death is paid for not at the end but at the beginning of life?

Near the end of the movie *Saving Private Ryan,* the dying captain, played by Tom Hanks, looks up into the face of Ryan, played by Matt Damon, and gasps one final order: "Earn this." For many men had just given their lives trying to find Ryan and send him home to safety, unlike his three brothers who had already died. Consider, also, the animal world. The females of several species of spiders eat their mates after copulation. By "willingly" dying the male ensures more time for his sperm to fertilize the female—the meal distracts her from moving on to another male immediately. These are just two examples of how death leads to life, each enticing us away from a narrow focus on a single individual's life to behold a wider world of connections between beings. And, as I intend to show, the stretching occurs in places one at first might not expect. With the enlarged view death appears not antagonistic to life but integral to it.

It's all a matter of scale and this book celebrates scale. Indeed, the key is to seek a much broader link between life and death than the usual sense of death as eventual arrest at the end of a single organism's life. It will take a book to elaborate the idea "death, thus life" across various realms, from society and eventually down to bacteria. My quest is to build a secular cosmology of death. From this we might gain an appreciation for the wondrous links between death and life, and thus nurture the opportunity we have to live as fully as possible in moment-to-moment awareness.

Who am I, as author? I'm not a therapist. I'm not a mortician. I'm not a hospice worker or even involved in the health professions. I don't sit at home dressed in black, watching the video *The Faces of Death.* I'm not a policeman, or war general, or writer of murder mysteries. I haven't enjoyed vampire movies since I was thirteen.

I am, simply put, a scientist. Earth biology—life on the planetary scale—is my trade.

For years I have harnessed computer models to help decipher how biologically essential elements travel within land, air, and sea. These building blocks of all creatures—carbon, nitrogen, phosphorus, and the ten or so other elements—wend in and out of organisms and swirl around the planet, all the while manifesting in a variety of chemical, molecular forms. Some of my studies have unraveled the reasons for flows of carbon dioxide between ocean and atmosphere. And I have peered back in time into the way the evolution of land plants changed the atmosphere hundreds of millions of years ago. I have also applied my knowledge about life and chemical cycles to help NASA design self-sufficient habitats for future space colonies on the Moon and Mars.

This work has allowed me to understand some aspects of the world that epitomize the concept "death, thus life." These aspects are especially profound because they are deeply ancestral to any sacrifice of war hero or spider. A wheel of death rolling continually into life propelled evolution along from the earliest bacteria through billions of years to the first human handprint on a cave wall. The key to the wheel, of course, is recycling.

Biological recycling is the worm that munches leaf litter into microscopic bits that are then further degraded by bacteria into nutrients that later can become tree leaves again. Death makes life. My favorite way to present this wheel is to actually put a number on it, to know exactly how much life death makes.

Imagine a world without recycling, where the small plants, tall trees, and marine algae all possess bodies so tough that nothing can digest them.

When they die, because they are indigestible, they will be buried as bodies in sediments, just as in our world small amounts of the dead slip through the food webs of worms and bacteria undigested. For subsequent generations of photosynthesizers, new nutrients must come directly from the Earth below, from rocks or volcanoes. As a result, photosynthesis, though still active, would be much diminished.

Looking at the element carbon we can know how reduced life would be. The current need for carbon from carbon dioxide by all photosynthesizers totals one hundred billion tons per year. Yet only a half billion tons per year is supplied as new carbon into the biosphere from rocks and volcanoes. Without recycling, global productivity would only be a half billion tons per year, a mere two-hundredth of its current value. Invert this number and we can say that in our real world recycling the dead increases all life two hundred times above what it would be without recycling. Death, thus two hundred times more life.

It is, of course, not death alone but the management of death by life that amplifies life. What happens at death to organisms is an integral part of the larger system—the biosphere with its internal wheels of elements. Most dead creatures are not just molded into the next organisms in the food chain. Dead individuals or parts of individuals are subsumed into a gigantic functioning system, as parts of creatures chemically transformed go into the globe-spanning fluids of air and water. For example, a dead leaf fallen to the forest floor will be consumed by dozens of species of soil detritus feeders, from worms to bacteria, who release some of the former carbon in the leaf into carbon dioxide in the atmosphere. Thus what happens to the dead amplifies not just any one other organism but to some extent all of life. In contemplating death we must attend to a very large picture indeed.

My excitement with these cycles is not just a case of scientific investigation but a fountain from which I take the edge off a thirst for immortality. I become other creatures through my atoms. I live again and again. I do not die.

But—is this all?

To Hamlet, the idea that we become food for worms only deepened his melancholy. Should it have? Assume my post-death carbon becomes carbon dioxide in the atmosphere, which then is pulled into the leaves of an oak tree to eventually take position in the tree's wood that lasts a hundred years. Is that a fate uplifting enough to carry me above the hell of knowing about my body's mortality? Is living on briefly in a bacterium's cell wall or perhaps even in its DNA a blessed truth from science that can support me through the tough times when I feel the truism of the bumper sticker that reads "Life's a bitch and then you die"? Can this vision—of recycled carbon, nitrogen, and so forth—as expansive as it is, answer my question: What is death?

Want to know the truth? No, it's not enough. Perhaps it is the primary truth about death. We might just have to bear it with a stiff upper lip. No one said that life was designed to make us grin back at its grim maws. But perhaps there is more, much more.

The more is not necessarily in a realm of heavenly angels, but right here on Earth. Perhaps there are broad aspects of "death, thus life" not revealed by the turning of the biogeochemical cycles.

————————

It is impossible for me to think back to winter of 1996–1997 without calling up the underlying horror of those months. Seclusion in a trailer a mile high in mountains, in a spectacular and remote corner of New Mexico, had seemed an ideal way to concentrate on writing a book about the global role of life in atmosphere, ocean, and soil. But after a sunny and productive summer and fall, I was sideswiped by a brush with death that came not like a raging dragon but like an insistent drumming in the dark.

The first signals seemed innocuous. The tip of my right thumb went numb. Then electrical zings began shooting along my arm at odd moments. A few weeks later I started waking at night with painful cramps in one hand or the other. Once I was jolted awake with my toes contorted and half of my face feeling like a wooden mask. Next my hands and feet started "falling asleep" in the middle of the day and would not wake.

Medical care was a problem. My regular doctor was two thousand miles away in New York. The nearest town was a two-hour drive, and its sole neurologist visited but once a week, weather permitting. So initially I hoped that my troubles would just go away on their own. Then, through serendipity, I discovered what I thought to be the cause of my infirmities: poisoning from carbon monoxide, emitted from a wall-mounted propane oven that had been activated just that winter after years of disuse. Monitoring with a meter, I discovered that airborne molecules of the odorless, invisible, deadly gas had at times been accumulating halfway to levels that could cause death in four hours.

Although it seemed I should have (literally and metaphorically) been able to breathe easy after this discovery, that was not to be. I stopped using the oven, yet I kept having what the neurologist (via phone) termed "relapses." I grew terrified as these "relapses" intensified. Soon I was barely able to write legibly. At night my brain would show me what was really in charge—the neurons—as trapped in the most inane obsessions I imagined myself, for example, peeling an apple for hours, unable to cease or think of anything else. Coordination faltering, I had to steady myself when walking, one small step at a time. Then my chest became the radiating center of body-filling pulsations, an uncontrollable drumming of rapid-fire vibratos that coursed along my arms and legs. Heartbeats pounded in my ears and set off reverberations in nerves elsewhere.

The mountain locale was so remote that once the mail was delayed for four days because of snow. In the middle of the madness, during slow, deliberate walks in the valley that sheltered my trailer, I began to make peace with myself, with my life now tethered to a nervous system going who knew where, perhaps to the final darkness. On one cold evening amble, with snow glossing the junipers and the shadows thickening, I relived my childhood and the entire pageant of my then forty-six years, coming to terms with the worst scenario my mind could conjure. I could be dying. Perhaps the carbon monoxide had triggered some deeper problem in my system, which was running me downhill. Perhaps I happened to contract some other illness, coincidentally, with the carbon monoxide exposure. After all, before my discovery of the carbon monoxide, an emergency room doctor had told me that my problems could stem from any number of rare but potentially fatal syndromes.

On that snowy eve alone in nature, I wept at the recalled beauty of snow-covered branches I had enjoyed as a child growing up in Buffalo. There we rivaled the Eskimos in our appreciation for the types of snow, its worthiness for snowballs, snowforts, snowmen, its effects on walking, sliding, and the inevitable shoveling. Then, as now, I appreciated the waning moments of daylight, when the snow purpled. For the first time in my life, I was being torn open by a moment of reminiscing as if I were summing it all up.

My ambition had always pushed me into more and more projects. Without children, books and technical results had been a way of leaving traces of myself in the world. And I was comforted by the knowledge that students carried on waves of my legacy into their own creative adventures. My first doctoral students were well on their way to professional success. And what about the children to whom I had taught science and math at a middle school for several years, or the art students who expressed to me how thrilled they were to be learning my principles of patterns?[1] I thought back to the mid-seventies, and my fumbling attempts to re-invent society, love, and work in the back-to-nature movement and to a passionate involvement with all sorts of other projects before I embarked on a science career. Then to friends I have known and loved. Such reflections during those pensive moments amidst the enfolding night, with the warm, yellow light from the trailer's windows visible some distance through the pine trees, helped me realize the fullness of my life. I became thankful. "I've reached an age greater than the average life expectancy for most of human history," I thought. "Who am I to think that I deserve seventy-plus years?"

Another month brought several more "relapses," and as the situation was clearly worsening, I desperately sought a solution. One morning I awoke, startled with a possibility. Testing the new idea, I put the carbon monoxide meter into the car, started the engine, turned on the heating fan, and, watching safely from the outside, saw the numbers surge into the danger zone. An exhaust leak! With every four-hour round trip to town I had been dosing myself with a second, independent source of carbon monoxide. A fatalist would think that someone was out to get him, but I am not a fatalist.

During an epiphany, the obvious, the banal, can flower into profundity. Sudden clarity allows you to witness a manifestation of Truth, and it changes your life. You can know that "all life forms are connected," for example, but to fully realize that fact, to experience it between yourself and everything around you, can awaken aspects of the world you never imagined. The same goes for other truisms, such as "life is short" or "we could die any time." My dark period of fear spun me face-to-face with these latter two, and the period is not over yet.

Even now as I am writing, if I halt all motion, I can sometimes feel my nerves pulsing, lightly there in the background, held at bay or masked by an anti-convulsant drug I take to counter presumed brain damage that never healed. The pulsing reminds me of the dangers in the world, of the nearness of death, which is no longer a grating insistence, just a gently rocking reminder.

When I talk about death in this book I mean the entire range of phenomena with which death is associated. This includes, often above all, the human awareness of death while we are alive. Why ask "What is death?" unless the answer is going to affect how we live? To think this way puts us in a league with the most profound minds of history, who saw how focusing upon death could propel them into more self-awareness.

For example, I have come to think of my problems somewhat as Montaigne, the sixteenth-century French philosopher of daily life, wrote about his urinary tract stones. His painful bouts came and went. Yet he claimed he was glad to be shaking hands with death once a month or so, because it was good for a man of his age (about the same as mine) to be brought home to meditations on mortality.[2]

One can think of Jesus, not only about how his death (and what came after) affects Christians, but as a teacher who wove death into several of his parables:

> *Jesus said, "There was a rich person who had a great deal*
> *of money. He said, 'I shall invest my money so that I may*
> *sow, reap, plant, and fill my storehouses with produce,*
> *that I may lack nothing.' These were the things he was*
> *thinking in his heart, but that very night he died."* [3]

Wake up, people! he is saying. Contemplate death and better your lives!

A world away a young prince learned this lesson firsthand. Prince Gautama had been sheltered in his father's palace of unsurpassed pleasure and ease all his days. But over a series of evenings he sneaked out by chariot into the real world and saw old age, sickness, and death. He also spied ascetic seekers, and right away abandoned palace luxury—to suffer, meditate, and eventually become the man known as Buddha. A vision of death and what he did with that vision was integral to his enlightened life.

As another angle on wisdom, consider shamanic traditions. These exist today, of course, and are also considered similar to practices ancestral to the formal world religions. Shamans are often "called" to their vocation by their own life-threatening ordeal of injury or illness.[4] Furthermore, during their apprenticeship, youthful shamans often have adventures that involve death during trances. They might have visions of dismemberment, boiling of their body parts, and separation of their individual bones. They often become skeletons. The lessons from such experiences turn out to be vital to their lives as practicing shamans.

But is just knowing deeply that you are going to die enough to help you develop fully as a human being? For some, certainly. But not for me. And not for you, the reader, if I may assume that's why you picked up this book. Some of us want more facts. Montaigne, Jesus, Buddha, and shamans—their insights and lives stir us to think wisely about the concept of death, but none offer a biological and psychological cosmology of death. Where will an encompassing vision of death come from? From science and, perhaps more importantly, from an all-embracing urge to delve into every niche of understanding about the mind, culture, the human body, trees, their cells, also bacteria, and the entire biosphere.

I have organized this book into three parts: Brain, Culture, and Biosphere.

In Part 1—"Brain"—the message is that mind and brain are one. This has burst like a supernova from discoveries in neurology and cognitive science during the recent decades. The implications are profound for the way we think about death, and thus how death affects our daily experience of ourselves.

Chapter 2, "The Three-Pound Miracle," discusses how knowledge about the brain's workings has made it difficult to believe in any disembodied existence after death, in the sense of mental beings without material embodiment. In particular, science has discovered the power of the cognitive unconscious, which operates by way of "organs" or subsystems within the brain. But what about consciousness itself? Is it also materially embodied? Apparently so, according to brain science.

In chapter 3, "We Live in Two Different Worlds," I acknowledge that many people still believe in what I call a dualistic universe. These folks say that matter and spirit are separate. They accept forms of mental existence that do not require matter and therefore can transcend death. As a scientist I cannot prove that such an otherworldy reality does not exist. But I challenge the dualists to produce evidence that can be verified through scientific, repeatable experiments. I look to the capacities of brains, such as hallucinations and other powers of the cognitive unconscious—coming from our brain's design as an anti-death organ—to explain experiences of spirits.

How we can employ our consciousness now in the service of facing awareness of mortality? This is the subject for chapter 4, "The Grateful Self." The question is whether consciousness has a general function in transforming our awareness of death into the impetus for a less fearful, more dynamic personal life. Brain science indicates that consciousness is a connector among many parts of unconscious cognition. If so, then how we think about mortality can help us develop cognitive stances and moral strengths useful in the ongoing predicament of facing our demise.

In Part 2—"Culture"—we move outward from the brain. Culture is the river within which we are ripples, a level of life broader than our individual selves. With its rituals and uses and ideas of death it helps shape us into who we are. How is "death, thus life" true on this larger scale of culture?

Chapter 5, "Nobody Just Dies," explores how social life is created around death. We explore here the universal functions of death rites in creating order from potential chaos and helping the loving feelings continue. Culture thus wraps itself around the death of individuals. We next consider the ancient, once all-important phenomenon of sacrifice. Again, social order surrounds death and is catalyzed by it. In sacrifices the uncontrolled—death itself—is precisely controlled. Sacrifices of individuals, either animals or humans, attempted to affect the larger order, often by communicating with the gods to benefit the whole of society.

In chapter 6, we examine some recent findings in social psychology called "terror management theory." Experiments have shown that awareness of death produces psychological reactions called "worldview defense" and "self-esteem enhancement." These are ways of coping with the primal collision of life: the urge to live and the knowledge we will die. The mechanisms of terror management operate unconsciously. But we can consciously take its findings to heart as we examine what kinds of attitudes we want to hold toward death in our daily lives.

Chapter 7 is titled "Death with Interconnected Dignity." I suggest the idea of the extended self as a way to express ourselves as individuals in a large social matrix that collectively deals with death. We need the funeral rituals because of their age-old functions, and we should be aware of the findings of terror management theory. But we can do better. The idea of death does not have to be relegated to days of mourning, or left to the whims of daily unconscious processes, but should be brought into consciousness.

Our search for how death makes life brings us finally to the biological roots of death, so Part 3—"Biosphere"—is about creatures. Why do humans die from old age, when some life forms, such as bacteria, seem not to?

In chapter 8, "Sex and Catastrophic Senescence," we marvel at creatures who die soon after copulation. For Pacific salmon and winged mayflies, life is fulfilled by a single, short period of mating. This seems like a waste of healthy adult bodies. But for them, death is the price they pay for the ability to reproduce. The logic is general, and can be extended to all creatures: death does not occur as means to open up space for the next generation, but happens as a consequence of exuberant living, whereby limited energies are devoted to a lifestyle that ensures the maximum production of offspring who carry the parental genes.

Chapter 9, "Lifestyle and Life Span," begins with the fact that some organisms, such as lobsters and bristlecone pine trees, exhibit negligible senescence, that is, they hardly age. What is the biological logic behind their enviable tradeoff between life and death? And what about the standard human life span of seventy or eighty years, in the absence of a tragic accident or untimely disease? Why didn't evolution provide us with more protracted lives?

Is the logic of death and life span basically the same on all living scales? Or does death have to be differently understood, depending on what level of life we examine? For example, what do we find on the microscopic level of cells? Chapter 10, "Little Deaths, Big Lives," examines the controlled, programmed deaths of cells within creatures such as trees and humans. These little deaths are regulated in two ways: both internally by suicide capabilities within all cells, and "socially" by signals between cells, thus integrating death into the life of the larger whole.

"Life and death at the smallest scale" is the subject of chapter 11. How does death work for single, complex cells that are free-living? What about the bacteria, famous for their ability to eternally divide and divide and never age?

The concluding chapter 12 proposes that we should keep in mind all the aspects of the bond between death and life at various scales, and draw upon these phenomena as sources of meditation and inspiration to help us more fully live in the present moment. In the end, all we have is the present. It is where we ultimately must work out and bring forth the creativity of our beings.

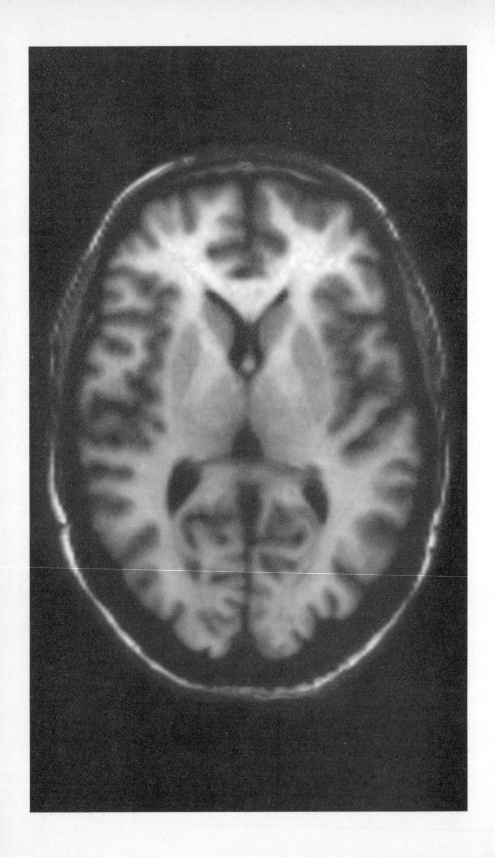

PART ONE

brain

THE
THREE-POUND
MIRACLE

A recent Christmas Eve party found me at my sister's home in New Jersey. After the train ride from New York, I needed the bathroom. I stepped into the small, dark room and started to close the door while still in darkness. Turning toward the wall, I simultaneously flipped up the light switch.

Unknown to me, behind my back, on a shelf at the opposite wall, was the ultimate in silly holiday gag items. As the room lit up with a screeching across the room—"Hiiii!"—I froze dead. Then began a familiar melody but in a raunchy, mechanical voice: "Jingle bells, jingle bells, jingle all . . ." My body racked with fright, I slowly turned around, and, as the song concluded with a shout—"Merry Christmas!"—finally saw the plastic apparition that had just scared me more successfully than the movie *Thirteen Ghosts* did when I was a kid.

The singing Christmas tree stood two feet high, topped by a red Santa cap. Several of its limbs formed gesturing eyebrows (in this case, eye-boughs), which beat up and down to expose bulging white eyes. Gruesome red lips were revealed by other undulating branches lower down (lip-sticks, perhaps?). It turned out that this monstrosity had been triggered by motion sensors. Thus I had to endure its jolly revival every time I made a substantial move in the room. So unnerving was it that over the next two days my nieces, ages eleven and thirteen, would always ask an aunt to accompany them or at least stand outside the partially open door whenever they used the bathroom. (I soldiered on alone.)

Earlier, on the train ride from Manhattan, I had been engrossed in a book that was, coincidentally, perfectly relevant to this experience. Written by neurobiologist Joseph LeDoux, *The Emotional Brain* describes a standard fear reaction, one that is basically universal for mammals.[5] I recognized it to a T: my freeze, the cautious turn, the stress hormones, the tactile stiffening of neck hair called piloerection. I had reacted with what LeDoux calls the "quick and dirty" processing pathway. Here's how it works. Sound waves enter the ears and are turned into electrical signals. These travel along neurons to a way-station in the sensory thalami (the sense-receiving portions of our two thalami), which are crucial egg-sized brain organs. The thalami will play a prominent role later in considering consciousness. From each it is a short run to the amygdala, of which we also have two, one on each side of the brain. Named after the Latin word for almond because of their size and shape, they sit just a little in front of and up from each ear canal, and inward a couple of inches from the skull. The amygdalas might be called our organs of fear.

Via the quick and dirty pathway the amygdala can then put your body on red-alert fear if the sound suggests danger. Given a loud, totally unexpected noise that detonates from behind—a possible threat to life—almost instantly the amygdala activates other brain regions responsible for stopping motion, elevating blood pressure, and releasing stress hormones that prepare the body for intense physical action (fight or flight). It's a fireworks of physical, hormonal, and nerve activity. And it's all without assistance from the conscious mind.

I can't really say *I* froze. I—as a conscious thinker—started participating in the situation a moment later when I—as a body—was already turning around. Furthermore, this moment ended at an eminently sensible time: soon enough for me to find out more information, yet slowly enough to keep me from inadvertently causing the potential danger to be jolted into more unpredictable action. By the time my analytical mind was kicking in—say in a second or two—the stress hormones had already pumped my senses into a super-focused state, had tensed my muscles, and had expanded the "present" by altering the perceived flow of time. This was all without "me" in control, a "me" who would likely only botch the operation.

My amygdala overreacted this time. The singing tree posed no danger. All the hormones awash in my body were to no avail. But when something makes a loud sound behind you—an explosion or an intruder, perhaps a cave bear?—it is better to be wrong some of the time than to pay the ultimate price for having underestimated the situation. And from a scientific standpoint, at least, it was an amazing experience. From a familial standpoint, of course, it was hilarious. Everyone was in stitches when they heard what happened, especially because they had been careful to drop hints to previous newcomers at the party that a surprise awaited them in the bathroom. But I had been told nothing. Amygdalic experiences aren't usually so amusing, unfortunately.

On my first summer stay in New Mexico I was still a greenhorn about the region's flora and fauna. During a mid-morning hike in a canyon, I left the streambed and climbed up along the bank, aiming for an overland path I'd spied on the map. Suddenly I heard a sizzle from under a rock right next to my feet. I rocketed off the ground, landing what felt to be ten feet away. Then I stared back at the huge rattlesnake under the rock, its tail still seemingly in my face and shaking furiously.

This was my first encounter with a rattlesnake, yet my reactions were automatic and correct. By the time I had any conscious, verbal thought, I had bounded and recovered. Hormones were flooding my body, which perhaps have lent to my memories of that moment an unreal quality: dreamlike yet simultaneously intense and alert, just like my recollections of the Christmas tree.

The amygdala can also learn. The next time I went into my sister's bathroom at the party, as foolish as it sounds, I was still a bit apprehensive. (As noted, my nieces had been even more conditioned to feel fear. Kids like to be frightened; perhaps they did not want to be deconditioned).

More crucially, during my years in New Mexico my amygdala helped me learn to stay away from any rock that offers a perfect shelter of cool shade for a snake on a scorching day. Walking within sight of such a habitat I feel just enough anxiety to shun the direct path past the rock. The anxiety rises before I even think much about the situation; the response is automatic, now burned into patterns of my neuronal pathways.

Before reading about the amygdala, I did not even know I possessed it. I cannot feel its presence in my brain. And yet it is there. It can be dissected and experimented with, allowing its functions revealed.

What would happen if we did not have the amygdala? Experiments on normal rats show the animals can be conditioned to flare up with a fear response to a bell, if first trained with a bell that is immediately followed by a nasty electrical shock to their sensitive feet. But such training cannot take place in rats without amygdalas.

And humans? Medical researchers at the University of Iowa have been able to study that question. Using a series of tests involving fear, they examined a woman who has serious lesions on both her right and left amygdalas.[6] She and others with normal amygdalas (for comparison) were shown slides of people with facial expressions of different emotions. All were asked to rank the faces as happy, surprised, afraid, angry, disgusted, or sad: a set of basic emotions. The damaged woman was well below the norm in her ability to register the fearful faces. She also had some trouble with the faces of surprise, which is not, in itself, a total surprise because of the obvious emotional kinship between surprise and fear. Furthermore, an excellent artist, she could draw detailed portraits of a typical face for all the emotions except fear; she explained that she did not know how a fearful face would look.

What I take from these experiments and my own experiences is a greater awe of our cognitive unconscious. (Sometimes I will call this simply the unconscious, but in modern cognitive science the term cognitive unconscious is used to emphasize that it consists of the entire system of mental operations that we are not conscious of, thus incorporating as only a subsystem Freud's unconscious and its role in repression.) Yes, I was aware of the fear and my responses once they happened to me. But I was not aware of the processes going on that made me freeze at the singing tree or leap from the snake. Had I no amygdala, nothing with which to process a fearful stimulus, then I wouldn't have reacted at all. I wouldn't have perceived the tree or snake as dangerous, and if, afterwards, it had been explained to me what I should have done, I might not even have understood why lack of reaction was inappropriate.

When they want to point out the powers of our unconscious, cognitive scientists look to everyday, simple speech. Somehow we talk and it all comes out fairly coherently. Yet so little is thought out beforehand. There are impulses of meaning, perhaps faint flashes of imagery in our thoughts, and then sentences tumble forth. And what makes these impulses, the flashes of intent that precede sentences, except the unconscious as well!

Perhaps you would like more conscious control over the cognitive unconscious, but we with our inadequate consciousness are far too puny for that. A striking proof of the limitations of consciousness has been provided by the cognitive scientist Bernard Baars, in an exercise that anyone can do.[7] Put down a list of numbers, such as 114723. Then try to keep them in mind while you read again the preceding paragraph with comprehension, pretending that you might be asked at any moment for this list of numbers. If you are like me, you can do either one or the other but not both. Keeping the list actively in mind turns the writing into gobbledygook. On the other hand, reading with attention to the meaning of the words pretty much dissolves the list of digits. Given the same reading task with only one or two digits to manage, the numbers can probably be maintained. But shouldering a slightly more complex list such as the one above maximizes our so-called working memory, from which consciousness draws in real time. In comparison to this small capacity, our unconscious could hold a vast library of volumes and analyze them.

The powers of the cognitive unconscious, such as those of the amygdalas, for example, are working all the time. Silently, thanklessly monitoring signals, always a step ahead of the conscious mind, their skilled efforts have served me in crucial situations. Every object I see or hear is being evaluated.

Is it a friend or dangerous foe that must immediately be reacted to? Imagine having to consciously listen carefully enough to each sound you hear to classify it as fearful or not. It could be done, perhaps. But that would be all you do. With the aid of the amygdalas and many other structures and pathways in the brain, the entirety of the unconscious rolls along as a thunderstorm, compared to that one raindrop of conscious activity.

———————

Gabriele Rico calls our brain "the three-pound miracle between our ears,"[8] and for all its hyperbole, that might still be an understatement.

Consider just the pathways for separating objects, say, a glass of water, into "what" and "where." From our eyes neuron circuits converge to the thalami and then diverge to regions in the rear of the head. From these primary vision processing regions, which are functionally devoted to a variety of tasks, such as creating color, size, and shape, two major paths feed the signals to more forward parts of the brain. The higher path goes to brain regions used in knowing where something is, which we need in reaching for that glass. The lower path, which goes into the amygdala as well as other brain regions, is used for the many tasks in which we need to know what an object is. Seems simple, but something in our brains performs this.

The differences in brain parts used for distinguishing types of objects is truly amazing. Knowing what an object is requires not just forward pathways of signals from the primary visual zones in the brain's rear but comparisons of the signals with what we know already. What do we know? That a hammer is a hammer and a dog a dog. Remarkably, brain scientists have found that the recognition of tools and animals is handled by different parts of the brain.

It had been previously known that one patient with brain damage might be deficient in naming animals when presented with pictures of creatures, while another patient had problems only with man-made objects. But now with modern techniques that record what regions of the brain are active when ordinary test subjects are performing specified tasks, the separate processing of animals and tools can be more precisely mapped.[9] During experiments, either task of naming pictures of animals or tools activated certain areas associated with speech. But the animal task also selectively stimulated a particular region on the brain's left side way in the back, a region involved with some of the earliest stages of visual processing. Naming tools selectively activated a region on the left side somewhat above the ear, a region known to also be activated when imagining hand movements. In addition the tool task fired up a region used to generate action words.

Brain-damaged patients have been found with problems producing vowels, but were normal with speaking consonants.[10] The inverse problem has been found in other patients. It thus appears that our brains separately process these two basic components of speech in different regions, at least in part. Furthermore, regarding sounds, a special brain region shows greater sensitivity to vocal sounds as compared with all other types of environmental sounds.[11] Many of these separations of function make sense because they reflect useful distinctions in our world. We need to know both what and where to take action. Special attunement to human voices seems very advantageous from an evolutionary viewpoint.

New technologies have allowed researchers to point out the active regions of subjects' brains and analyze even more complex mental activities. One recent example involved subjects who kept a main goal in mind while performing tasks toward secondary goals.[12] The experiments showed that certain regions in the front of the brain were active only then. These regions were not engaged with only either main or secondary goals performed by the subjects (who sought patterns, in this case, in sequences of alphabet letters). Researchers claimed that mental activity with both main and secondary goals alone is the essence of planning and reasoning, which perhaps is made possible by certain regions of our brains.

What about memory, such as that of personal experiences? A centralized part of the brain, shaped like a seahorse and called the hippocampus, is vital for this. In one case, a man whose hippocampus was damaged as a result of encephalitis could recall his neighbor as a child but, after the injury, nothing about where he currently lived.[13]

What about this thing called the self? Is it just the integration of everything in the brain? Or is the self specialized within the brain? Of course this all depends on what is meant by the self, and in the words of one brain research group, "there is no coherent body of knowledge that comprises a cognitive neuroscience of self."[14] I have examined endless pages from neurobiologists and philosophers who are not even able to agree on how to define the self. Neither can anyone be definite about the terms "mind" or "consciousness." All these—"self," "mind," "consciousness"—are employed in various ways by various people at various times. The terms more or less overlap, depending on speaker and circumstances. Yet we can say they all refer to something intrinsic to our brains. They are words we use to point out to ourselves, in our inner dialogues, something quintessential about who we are and how we operate in daily existence.

Despite difficulties with definitions, specific investigations have begun to shed light on aspects of the self. The above-noted research group claims, for example, that a variety of "self-processes," such as autobiographical memory and evaluating whether a picture of a face is of oneself or not, occur preferentially on the right side of the brain.

Some key experiments were conducted with patients who were undergoing anesthetization of either the left or right sides of their brains.[15] Surgeons needed to know about each patient's degree of hemispheric dominance prior to surgery to treat epilepsy. Brain researchers used this opportunity to test for the neural location of self-recognition. They showed pictures of faces to the half-anesthetized patients. These faces had been computer-generated as morphs between a subject's own face and a famous face. A woman might see a blended face that was halfway hers in its features and halfway that of Marilyn Monroe. Upon recovery from the anesthesia, the same patients were shown individual pictures of both themselves and the famous person, and they had to choose which one they thought they had seen earlier when half their brain was numbed. Patients who saw the morphed face while the right side was numb picked the famous face as the one they had seen. Patients recovering from left-brain anesthesia picked their own face as the one they saw during the numbing. After these experiments and a review of other findings, the researchers concluded, "It is conceivable that a right-hemisphere network gives rise to self-awareness, which may be a hallmark of higher-order consciousness."

No doubt the self is a complicated psychological concept, encompassing both conscious and unconscious parts of our daily existence. But we have little trouble pointing to ongoing daily experience of consciousness: the contents of our moment-to-moment "field" of awareness, which encompasses both objects of our senses and the inner feelings, imaginings, and silent dialogues that take place in our heads. To a large degree consciousness is us, and inroads here as well have been made by neuroscientists that indicate consciousness is surely brain-based.

Referring to consciousness, neurobiologists Gerald Edelman and Giulio Tononi state, "On the basis of studies of lesions and stimulation, we are confident that, for example, the activity of certain brain regions, such as the cerebral cortex and thalamus, is more important than the activity of other regions."[16] A standard illustration of this is the fact that if half of the brain is removed, which is done in some desperate situations of disease, the recovering person might experience cognitive problems but will still possess consciousness. In contrast, remove a couple key areas of brain and consciousness will be lost, even though the person remains alive.

Bernard Baars, a cognitive scientist, has summarized some of these key parts of the brain, which specifically function to make consciousness.[17] One is the reticular activating system, or simply reticular formation. Located in the brain stem, it is shaped like a narrow, squat funnel under the brain's somewhat slumped sphere. According to researchers Blair Turner and Margaret Knapp, "The function of this system is associated with arousal, and its destruction produces permanent coma."[18] This arousal is not the kind encouraged in you by an eager lover. The term in neurobiology refers to the fact that the reticular formation makes the rest of the brain sensitive and awake. Remove or damage the reticular formation and consciousness ends.

Baars points to another important brain organ, the thalamus. Each of the two thalami is an egg-sized "mini-brain within a brain," located about an inch above the reticular formation. We have already seen how each transfers signals from the ear to the nearby amygdala. In fact, each is a central relay station. The thalamic eggs connect with virtually every portion of the brain that involves sensations, motor commands, attention, and other aspects of cognition.

Scoop out a couple egg-sized spoonfuls from the top of the brain and a person might suffer a few problems, perhaps in willing some part of the body to move. But full consciousness remains. On the other hand, take out or damage the two thalami, which total only about five percent of the brain, and consciousness collapses. A person would then enter what is called a persistent vegetative state. An example was Karen Ann Quinlan, the young woman whose lack of consciousness while she remained alive for nine years in a persistent vegetative state spurred the development of many new legal and medical guidelines. Upon death, when her brain was autopsied, the most significant damage was found in her thalami.[19]

It might even turn out that not the full thalamus but crucial parts of it are the key. Within each thalamus, in between folds of tissue called laminae, are several structures the size of pencil erasers: thalamic intralaminar nuclei. So pervasive are the connections between these nuclei and the brain's surrounding, large cortex that the special nuclei have been compared to a spider in its web. Referring to the nuclei, Baars states: "Aside from the reticular formation, this is the only part of the brain that seems indispensable for waking consciousness."[20]

Consciousness requires contents that include sensations, thoughts, images, and feelings, and this complexity of components implies the presence of a network of processes within the brain, not just any specific structural part, as important as certain parts might be. For example, Edelman and Tononi point to how the thalami and the cortex are connected. Structurally, the cortex is like a fat baseball glove that surrounds the thalamic eggs. In the words of these researchers, "Consciousness is a dynamic property of a special kind of morphology—the reentrant meshwork of the thalamocortical system—as it interacts with the environment."[21] Their system refers to thalamus, cortex, and internal connections, which the researchers call "reentrant." This term means that wherever a region in the cortex connects to a region in the thalamus, then the same region of the thalamus connects directly back to the region of origin in the cortex.

The special morphology of reentry produces the ongoing series of unique states we experience as our conscious minds, an integration that Edelman and Tononi call the "dynamic core." The dynamic core is like a tangled network of central springs with more peripheral springs. Disturbing the peripheral springs can set the entire central core vibrating. The dynamic core both changes all the time and maintains a certain constancy, somewhat like a flame, flickering in endless variation yet still located at the wick of the candle. Consciousness might be seen as a vibrating flame in the brain.

And death would be its snuffer. When the brain's activity ceases, consciousness must also cease.

Here is the rub: If our brain is the source of self, the main arbiter of our senses, the pilot for our bodies when we have no time to think for ourselves, the source of daily consciousness, then how could we live without it? How could we have a self once the brain has been denied us by death? How could we sense the glories of heaven or the pains of hell without a brain to process them? How could we enjoy any mental capacity at all? Thus with all the discoveries about the brain flooding the technical journals, it is clear we cannot possess a soul. After death, there can be no reincarnation of self because the self is forever tethered to the body it was developed within after birth; it cannot go somewhere else and wait to enter a new body some time in the future. There can be no eternity in any heaven or hell, no purgatory, no bardo states. If you want to find nirvana, then you have to look for it here.

WE LIVE IN
TWO DIFFERENT
WORLDS

What would we be when we die, without the living brain? We could not talk, or distinguish tools and animals, or reach for a glass in some otherworldly place. Life in heaven is often portrayed as an idealized version of life here in the material world. In heaven, people communicate to each other, they often have something like bodies made of spirit, and they see each other. But how could people in heaven recognize each other without a brain that has special circuits for processing individual faces? How could they talk without a brain to store, select, and order words? After death we could neither plan nor reason. We would be unable to form new memories. So what would we be? Following this line of logic, I cannot help but believe that we terminate as individual psychological beings at the moment of bodily death.

Some of you may feel the same way, and I'm sure some of you do not. To distinguish these two camps I'd like to propose some terms. I'll use "monists" for those who believe that there is a oneness of brain and self, that together they form a single system, indivisible by death. "Dualists," on the other hand, see the brain and at least some aspect of the mind as independent entities detached at death. This aspect is sometimes called the soul, personal spirit, interpersonal spirit, or atman—hundreds of names could be found in the anthropological literature of world cultures. Whatever the word, dualists accept both this plane of earthly reality and another very different plane—of astral existence, nonmaterial spiritual beings, angels, formerly alive people, disembodied souls. Whereas monists accept just the former.

Dualists could point to the difficulties science has in explaining consciousness. This is true. Until recently, it was not even considered scientific to investigate consciousness because such studies always seem to depend ultimately on a large dose of subjectivity: namely, data must come from what people report they experience. How can these reports be verified? By studying the effects of different types of brain damage and considering the data from the latest generations of scanning technologies, a foundation for a science of consciousness is being laid, but there is still a long way to go. Antonio Damasio, one of the looming figures in the science of consciousness, has recently written, "Not only is it true that we have not yet exhausted the possibilities of explaining consciousness in neuroscientific terms, but it is also true that we have barely scratched the surface of neuroscience in terms of such an attempt."[22] Yet scratching the surface in this case is quite the something. And what we do know indicates that consciousness is brain-based, strengthening the case for monism.

For many dualists the logical bottom line is not any counter to the findings of neurobiology but the fact that they often base their beliefs on personal experience. In working on this book, I have found there's nothing like the topic of death to take conversations with friends and relatives straight to the heart of beliefs about reality, to bring out partial, if not full-fledged, dual-ists. One professor of English told me that half the time he thinks he will end when his brain dies, the other half he thinks there is some type of existence after death, something far more than has been recognized by science. Then he told me about a number of odd experiences he had been through which con-flicted him.

Now, I am no stranger to strange events. And I am somewhat in awe at how many people have had experiences that lead them to believe in out-of-body travel, telepathy, even contact with the dead.[23] But I have never had anything happen, even in my "mystical" experiences, that absolutely denies an explanation either by statistics or by unusual and rare brain events. In fact I support the quest for such experiences because I have learned from them, with new insights about my place in life and how to plan my future, to men-tion just a few positive results. But as important as these have been, I still believe I am playing by way of my brain. No brain, no play.

So how does a monist explain what dualists are experiencing? Should their experiences convince me the dualist position is correct?

———————

My sister Kris is the dualist I've talked to at greatest length about such mat-ters. She is a hypnotherapist in Minnesota.[24] She helps people and often has extraordinary results. Under hypnosis a woman recalled where she put a valu-able ring that she hid and then forgot. Clients have quit smoking after one session. The sessions typically involve guided imagery. Kris herself is a power-ful visualizer. Using her soothing voice, she helps others to travel in their imaginations and reprogram themselves in desired directions.

Kris has had some amazing experiences herself, none any less remarkable than those typical of esoteric or New Age literature. She is my sister and so I have had access to her for in-depth talk. I also trust her. She is not making money off her own remarkable moments, and she is personally questioning what life, consciousness, and life after death are. I enjoy testing her with every technical brain insight I can muster. I am truly interested myself, as a seeker and as a scientist, so I take her moments to heart and listen to her, as well as debate.

She has visitations. As a child, not uncommonly she would awaken at night to apparitions in her bedroom. Sometimes these spirits sat on her bed and talked with her. Her visions still occur. One recently involved a luminous female being, again in the middle of the night. The wonderful being of light told Kris something important about her life. To hear her speak of the vividness of the spirit guide that night!

Kris told me about a time she was visiting a woman friend at the friend's remote cabin on a lake in western Minnesota. The friend's father had died several months earlier. Kris had known him. It was Kris's first time at the lake property. Soon after having arrived, she was walking a path, by herself, up to the cabin. Suddenly there he was—the dead father—appearing in broad daylight before her on the path. He greeted Kris and welcomed her to the property, which had been a special, sacred retreat for him. Kris later told her friend about seeing the father and pointed out where the visitation took place. Her friend, not at all surprised, then showed Kris that right nearby was the tree where the father's ashes had been placed. Kris had not even known the family had placed ashes on this cabin property. I asked Kris if the dead man had looked as vividly solid as ordinary people do, or more like a blurry apparition. She said the latter.

How should I interpret Kris's experience? Do I take it as evidence of the afterlife existence of an individual? Should I consider giving up monism for dualism?

I will rely on a general explanation for such mystical experiences. With apologies to any injustice I might do to different types of dualism, I here lump together the mind-to-mind communication called telepathy; out-of-body experiences in which one travels somewhere and sees things in the world without usual physical limitations; clairvoyance which includes knowing things that no one knows without the ordinary senses, such as locating where a murder victim is buried; and other tenets of dualism, such as communicating with angels that give advice during their contact with the dead. All these experiences presume a reality quite different from the one ordinarily accepted by science, which includes atoms, electromagnetic waves, gravity, and the structures built from these, a reality whereby already known, basic, physical parts become elaborated upward in scale in very complex ways, such as in bodies and brains, for example.

To start, we must admit that memory can be biased. How accurately does Kris recall her experience? If a person is predisposed to believe a palm reader, the person tends to emphasize the reader's hits and ignore the misses. Memory interprets and often leans toward a desired outcome. What do we desire? Psychological comfort, for one. And for some people there is certainly a good dose of comfort in dualism because it offers the possibility of immortality on another plane of existence. Therefore, given our desire for solace, it would be all too human to skew memories of experiences by shaping them in ways that support dualism.

This skewing of memory occurs in the cognitive unconscious, which then brings the altered memory into consciousness. What else does the unconscious do? Mainly, it's always figuring situations out. This figuring out is what we as humans excel at. Furthermore, people figure things out in a variety of ways.

The theoretical key here is from the work of Harvard psychologist Howard Gardner.[25] It's called the theory of multiple intelligences. Gardner has developed evidence of a small number of fairly independent kinds of intelligence, somewhat like modules with different capabilities. People are not equal in all the types. One type, for example, is what I am good at: logical-mathematical. But there are others: linguistic, musical, spatial, bodily-kinesthetic, and two "personal" intelligences, the introspective and the social. This list is not arbitrary. Gardner gathers proof for each type from two principal lines of evidence. First, people can excel and even reach a genius level in expressing one of the intelligences, while being much more normal in the others. Brilliant composers (Mozart), mathematicians (von Neumann), and dancers (Duncan) are premier examples, respectively, of musical, logical-mathematical, and bodily-kinesthetic types of intelligence. Second, brain damage can cause people to become defective in specific types of intelligence. We have earlier seen how damage can cause anomalies in, for instance, naming animals but not tools. As Gardner points out, a vulnerability to disruption of basic types of intelligence indicates functionally unique brain systems. People in general are mixtures of all the intelligences; people excel in a few, are normal in a few, and are low in a few (not being able to carry a tune, for example).

Social intelligence is the ability to read other people, their emotions and intents as revealed in their faces, their unconscious motives that leak through in their speech, their feelings in their posture and bodily attitudes. The social intelligence would possibly be, in archeologist Stephen Mithen's theory of the evolution of the human mind, the oldest of the specialized intelligences, inherited from ape-like ancestors and a vital necessity for life in highly cooperative and competitive social groups.[26]

When instead the spatial (or, in other words, visual) intelligence blossoms more colorfully than any of the others, we have the makings of an artist, one who like Renoir could respond to olive groves such that "the sky that plays across them is enough to drive you mad."[27] To this intoxicated, ultrasensitive response to the visual world, the artist adds his or her own visions. The inner eye of one's own imagination is extraordinarily strong.

So through the lens of multiple intelligences I will consider some of Kris's uncanny visions, which she sometimes has during her hypnotherapy sessions. While a client is hypnotized, Kris has seen a visitation who interacts with the hypnotized person. It might be a mother behaving as she did when the client was young, or a teacher, again in a scene that Kris felt was a replay from the client's youth. She has sometimes been accurate to a degree that the vision allows her to reveal a deep, hidden source of the client's trouble and catalyze much healing right then. What happened? Did Kris receive signals telepathically from the client's subconscious? Did she see the past reenacted because it still exists somewhere on the astral plane? Kris definitely interprets these visions as support for her dualist outlook.

I think her cognitive unconscious figured something out and gave her an answer in the form of a hallucination. She combined a brilliant degree of social intelligence with her visual intelligence.

What are hallucinations? In a sense all of us often have mild versions. They occur every time we visualize an image or talk silently to ourselves and hear that wondrous patter of our own inner voices. In these mild versions we know that inner images or words are "inner." We do not project them out into the world. But consider the following example.

My former graduate assistant, Aaron Krochmal, was once terribly sick in a fever, probably due to some strange bug he picked up while doing his field research (for another professor) in the swamps of northern New York State. Aaron, sick in bed during the first evening's bout of illness, had a long conversation with his father. The family dog, too, was in the room. The next day when his father again came into the bedroom, Aaron started up with the previous evening's conversation. But his father had not had any conversation with Aaron, had not even entered the room that evening, in deference to letting Aaron sleep undisturbed and better recover. The dog had not been present either. Aaron told me it was exactly like seeing a real person and real dog. His hallucinated father even spoke like his real father, with the same voice and phrases and personality. Aaron recognized his hallucination as such, but only the next morning, when his Dad couldn't confirm the visit independently.

The problems left in my nervous system from the carbon monoxide damage have given me sensory hallucinations that clearly point out the potential problem of distinguishing outside from inside. Once, before I started treatment with the anti-seizure medication, I was standing in the Houston airport, waiting for a plane to depart. Suddenly I felt a subway rumbling under the floor, just like in some buildings in New York. I even said, Oh, the subway just passed. Then I realized: Wait a minute, I'm standing on a huge mass of concrete floor on the outskirts of Houston, which doesn't have a subway. I had interpreted my internal sensations as external events.

One brain discovery that gives insight into the process of misinterpreting internally generated sights and sounds is the fact that such internal creations originate in areas of the brain that process ordinary vision and hearing. This creates a potential issue for us. Cognitive neuroscientists Chris Firth and Ray Dolan note that this dual functioning of these brain regions "raises the problem of how we know whether our experience derives from mental imagery or from something happening in the outside world."[28] What if internally triggered patterns occur at the same time and in the same brain sites as externally triggered patterns? With internally generated and externally generated patterns in the same brain sites, the internally triggered events could be interpreted as coming from the external world.

Consider some details of vision processing. Seeing a shoe is not just a matter of seeing a shoe. The shoe must be interpreted. It turns out that the first areas in the back of the brain to receive signals from the eyes do not just process signals in various ways and then shunt their findings forward to "higher" brain regions. Rather, patterns from the more frontal, "higher" regions are simultaneously flowing toward those early, back-of-brain, initial processing sites, thus providing interpretive patterns from higher regions to guide the endeavors of lower regions. In short, the pathway for object recognition is not a one-way traffic flow. Instead, cognitive operations from the more frontal regions of the brain seem to be coursing back along the forward pathway of visual analysis. The two flows influence each other. As visual processing is flowing forward, memory linked with interpretive reasoning is flowing backward. The memory acts as a test model during acts of recognition.[29]

This concept of counterflows along pathways in the brain is a key principle, highlighted by virtually all neuroscientists. Recognition is thus the complex merging of sensation and interpretation. Interpretation occurs even when distinguishing an apple from an orange. But suppose that in our interpretation of seeing an apple, we also imagine a person picking up the apple. Is it possible that a strong visualizer could see the person picking up the apple and think the imagined person was external? If I do this exercise, the imagined apple eater is vague and I clearly know it is imagined. But what if the imagined apple eater occurs in the brain of a brilliant visualizer, one who even has a history of interpreting visions as coming from the external world? The brain's anatomy provides the potential to conflate and thus confuse externally and internally generated sensations.

So I think Kris hallucinates in a way that derives from the activity of her cognitive unconscious as it figures out some situation by means beyond her conscious control or bidding. (She says such events usually happen of their own accord, but do occur during times of her life in which she feels more "open.") Kris, in my opinion, does not read her clients' minds using some as-yet-undiscovered psychic wavelength of transmission. Rather she reads minds using the clients' body language, which is then translated by her unconscious as a vision in conscious awareness. She has combined a sensitive, highly tuned social intelligence that intuits aspects of a client's past and projects possible answers in the form of hallucinations.

The visitations Kris sees when she is alone are easier, of course, to explain as hallucinations. But they, too, are intriguing, because they come from her unconscious, as fears and hopes of specific situations, even though the exact meaning is not always obvious to her. In experiences where people see and/or hear dead relatives, we could have other moments during which the unconscious minds of the living generate internal visions or voices, then project them as external, as part of the healing process. Moments of trauma, such as death, would more easily bring about such bursts of exceptionally strong imagery into consciousness. The overall experience may be likened to having a dream in the middle of being awake.

I will admit that I don't have an easy explanation for Kris's vision of her friend's dead father. Certainly being at his country retreat and knowing of his death could have triggered Kris into enacting a reassuring vision. The fact that the vision occurred near the site where the family placed the ashes was luck, a bonus from chance in that particular situation that made the dualist view even stronger.

If the vision had been elsewhere on the property it still would have been remarkable, but more obviously interpretable as a hallucination. Andy Neher is professor emeritus and former chair of psychology at Cabrillo College, on the California coast. This is near Santa Cruz, so Neher has been in the middle of parapsychology "New-Age-ville." He has heard it all. And he has investigated a lot of it. His book, *The Psychology of Transcendental Experience,* is a classic. In it he documents the various ways that people use their brains to open themselves up to unusual experiences, which often lend support to a dualist viewpoint. These ways include trances, hallucinations, visual and auditory illusions (the ways the senses can be fooled), group dynamics that lead to instilling false beliefs in participants. It is not that everything he has run across can be explained. But, over the years, the lack of respectable evidence of the afterlife and paranormal has left Neher skeptical. As a scientist he is open to verifiable new findings that would change his mind. And we all must remember that absence of evidence is not the same as evidence of absence. As valuable, however, as he thinks these unusual experiences can be for people of all ages and walks of life, he does not feel that the brain is absent as a player or that a person would have these experiences, let alone the ongoing experience of daily life, without the brain.

———————

More skepticism comes from one of the world's authorities on parapsychology, Susan Blackmore, Ph.D. and Reader in Psychology at the University of the West of England. She has investigated near-death experiences, out-of-body experiences, telepathy, clairvoyance, contact with the dead—you name it and she has sought it out in the action pits with the people claiming to have these abilities, and has written books on the topics. In her updated afterword to her book *Beyond the Body: An Investigation of Out-of-the-Body Experiences,* she explains how she tested some self-proclaimed OBEers in England.[30] She posted targets of words and objects on the inside of her door, at random. Then the OBEers were supposed to visit in their OBE and see what was there. It was a failure.

Blackmore recently described her early investigations.

> *Thirty years ago I had the dramatic out-of-body experience that convinced me of the reality of psychic phenomena—and launched me on a crusade to show all those closed-minded scientists that consciousness could reach beyond the body and death was not the end. Just a few years of careful experiments changed all that. I found no psychic phenomena—only wishful thinking, self-deception, experimental error, and even an occasional fraud. I became a skeptic.*[31]

She continued, year after year, testing a variety of extraordinary claims. Finally, after thirty years, she gave up altogether. If there is something parapsychological out there, in my opinion Blackmore would have bagged it.

We can look farther back in time for serious research into parapsychology that failed. About a hundred years ago, the newly organized American Society for Psychical Research convinced renowned astronomer Simon Newcomb to take the presidency. Scientists in the society hoped to establish a "systematic study of phenomena such as telepathy." "Never one to investigate a topic halfheartedly, Newcomb threw himself into the thick of research on paranormal events, poring over the literature and attempting to witness occurrences firsthand." But after his investigations, which included "mediums and other masters of the paranormal" in various U.S. cities, he found nothing. In his presidential address he decried those willing to infer "new laws of mental action without being able to replicate the relevant phenomena."[32] Despite his dissenting view he was reelected and afterwards remained as a member of the governing council, trying to convince others that "psychical research was a scientific dead end."

As I look across at those on the other side, those who infer new laws of mental action, here is one of my major problems. For all the accounts of pastlives, we are faced with a shocking lack of any hard-core, verifiable, replicable, scientific findings regarding such purported phenomena. This is not for want of trying. But the fact is that we have not progressed in bringing any scientific light to these phenomena. Society still has what it had a hundred years ago: throngs of believers with personal anecdotes.

Certainly science as a process is, in my opinion, the supreme example of being tentative, open to new findings. New discoveries are always being made. It would be futile hubris to state that science knows all. So some dualists would say that science hasn't yet found the equivalent to Galileo's telescope, able to provide verifiable new data about the realm of angels or the wavelengths of telepathy or vibrations upon which the dead continue without their bodies. Is the study of the afterlife waiting for its Galileo and telescope to open up new visions for all to behold?

There is a problem with this line of reasoning. We have a world in which countless numbers already claim to possess the psychic telescope, to have telepathy, clairvoyance, and the ability to communicate with the dead. Why have these phenomenal people not been brought into the forefront of science? All it would take would be one genuine psychic to put on those half ping-pong balls over both eyes, have another psychic in a remote and sealed room transmit pictures, then have the first psychic successfully receive them. Then let the pair travel the wide world performing in all the labs, under various degrees of rigor and experimental control, and have it be true, true, true. The headlines would trumpet the finding everywhere that television or printed word reaches. Because the psychics want to spread the message of their powers, including mind-to-mind transmission and contact with the dead, why don't they just do it—prove it once and for all?

After Galileo turned the telescope on the heavens, anyone could build the device and see for himself or herself. So we have the following situation. Millions can do what no scientist can do in a laboratory experiment that others could duplicate, and which could pass the peer-review quality assurance so essential to scientific progress. It's as if in the days of Galileo millions said they could put a glass to their eyes and see sunspots—but no scientist could. What a weird, split world that would be. Yet, with regard to telepathy and contact with the dead, that seems to be the world we have. How is this possible?

Believers in the afterlife and paranormal could say the fault is not with the replicability but with the science profession. They could say that scientists are stuck behind blinders put on them by their profession so that these phenomena won't be investigated. Or perhaps positive scientific results about the paranormal are not being allowed into the top journals by the gatekeeping editors and reviewers of the science priesthood, who forbid any such knowledge to pass with their blessing just as the priests of old tried to suppress the nascent findings of telescope science. But in the reality of professional science that I know so well, scientists are hungry for big discoveries, champing at their laboratory bits to have a chance at the fame that would reverberate across the decades. And such is what would come from any simple, reproducible demonstration of disembodied intelligences, telepathy, or contact with the dead. Real evidence for any of these would win the Nobel Prize of Nobel Prizes. But from the vantage point of scientific investigation, there is nothing but failures, flawed experiments, botched statistics, and, as a fact of history, fakes.

The future of science is likely to help us reach consensus about whether the monist or dualist view is correct. Right now the monist view is supported by both the lack of laboratory evidence for a dualistic world and the lack of any explanation for how the dualist view could be true, given what is known about the brain.

To repeat, it is always important to remember that absence of proof is not proof of absence. Certainly modern physics has its share of odd phenomena and theories to allow for something more radical to emerge about the nature of mind than we can even yet imagine. Take, for example, the superstring theory of ten-dimensional objects of which six dimensions are hidden, giving us our three of space and one of time. Or consider the quantum-world finding of nonlocal interaction. But I'm not holding my breath. I'm going to follow the monist path and move along in developing a personal philosophy of how to live life in the presence of death as final extinction of myself. The question now is: Where will this path lead?

THE
GRATEFUL
SELF

What does the monist view ultimately offer? A bunch of neurons? That we are only equivalent to a cantaloupe's volume of specially sensitized cells? However magical their trillions of firefly flashes of activity seem to be, however much their mysteries are being unraveled by science, however energized we may feel by imaging the electrical, chemical pulses among their networks, it all still feels like a desperate act to try and find meaning within them. Taking on the task of finding meaning in neurons can make one feel like the character in Edvard Munch's famous painting of an anguished, despairing face. Do monists have anything to offer other than The Scream?

We must be careful not to equate the fact that we would not exist without the material brain with saying we are only the brain. We are an organizational property of the brain—and entire nervous system and more—a property of the whole system. In this way we, as mental human entities, are similar to the property we call "life" as it arises from the complex interactions of the parts of a cell. We experience some of the neural orchestration as our magnificent moment-to-moment existence: consciousness. It is in consciousness that we should seek for meaning in how the concept of death can enhance life.

How should we consider the thought of death in the context of what is known about consciousness? Is death a thought that pollutes? A dirty secret we should cover up? An idea we should smash like a false idol or run from like poison gas? Perhaps we should not make it the focus of consciousness because of its capability to produce fear.

This fear is not a fear of something concrete, like a singing Christmas tree or a rattlesnake. With them the danger either is or isn't there, and so the fear comes and goes as a useful function of the brain, prompting us if necessary to perform in ways that avoid the danger. Instead, the thought of death can be there all the time. It's the ultimate predator, with canines ready to bite and a maw ready to swallow us, if not now then someday. If dangerous and real sights and sounds can throw our amygdalas into overdrive, then what does the idea of death do? Could it make our amygdalas cringe in response at every moment?

Should we instead sublimate it into a variety of phobias and fear disorders that sometimes involve death, such as in panic attacks, where the victim feels as if he or she is dying? If the thought of death creates a life of dread and fear, is this the life we want? If not, then maybe we should keep death away from consciousness. Perhaps the only good thought of death is an unconscious thought, and thus, as far as we conscious beings are concerned, no thought at all. Of course, according to classic Freudian psychology, repressing the thought of death might be equally unwise. Because the unconscious is so powerful, the repressed thought would still be there in the deep mind, haunting us.

Many wisdom traditions around the world provide us with another option. They tell us to not repress death but to look at it, to make it conscious as a way toward personal growth and development. As the examples (from Montaigne, Jesus, Buddha, and shamanism) in the introduction show, the idea of death can create a more enriched life. We looked earlier at how Buddha left his life of ease, a move triggered in part by seeing a dead person. Now let's hear Buddha at the end of his own life.

Nearly eighty years old, Buddha called his monks over to his deathbed for some last words. He first spoke about the continuance of his message not through his moribund body but through his teachings. He then gave his famous final lesson: "I now impress it upon you, decay is inherent in all composite things; work out your salvation with diligence!"[33] We all are going to die—so keep this fact in mind and pay attention to your lives now.

Perhaps any benefit of making conscious the inevitability of death is like the benefit derived from proper composting of kitchen scraps. If kept compacted in a pile, the scraps begin to smell and can become toxic, as anaerobic bacteria colonies get to work. Instead, the scraps need to be aerated and tended mindfully, then they can be spread in the garden to provide nutrients for plants. Regarding mortality, tend it and turn it and eventually mix it into other parts of the unconscious so it can be spread throughout, possibly allowing powerful benefits of life to emerge. And more—keep the fact in mind, Buddha said, be conscious. That's how nature built our brains to be used.

The thought of death can be imaged as a seed, which grows from being planted in the mind. But into what? We could offer a possibility by interpreting a famous parable of Jesus here. In this he referred, as he often did, to heaven's imperial rule. No one can know for sure what Jesus meant. But the phrase can be taken to apply to the here and now. In other words, in my opinion, he was describing an advanced state of being in life. Heaven's imperial rule, he said, is "like a mustard seed. It's the smallest of all seeds, but when it falls on prepared soil, it produces a large plant and becomes a shelter for birds of the sky."[34] This metaphor could apply to any emotionally laden ideas that help one grow in a beneficial way. It could apply to the concept of death: plant it in consciousness and it will help you as it spreads to become a haven for developing the way to live.

And yet—despite these metaphors and promised benefits, the earlier question remains. How can we be sure that it's not fear that will grow from making the idea of death more conscious? It's all very well to say the seed will grow, but what is the species of the plant? What if we achieve only thorns in the mind, not sheltering fronds?

For example, Eskimo shamans of a certain tradition, through physical and mental ordeals, hone the ability to see themselves as skeletons.[35] This practice may seem terrifying. Yet we know that such people are revered within their societies as evolved beings, able to help others. What exactly is going on?

Surely the basic starting point is that we as metabolic beings will likely experience fear when contemplating death, because this idea is a biological threat, and we have evolved to become fearful and take evasive action against threats. That is probably why we so easily, naturally, repress thoughts of our mortality. Fear is a biological reality. Fear is natural in this case. If we as psychological beings face the challenge of meeting a concept of death that we sometimes call up inadvertently, feel tentatively, and often hide from, how much more challenging it would be to face a concept of death that we consciously emphasize and amplify! Now consider the way that biological threats of death from the environment helped organisms evolve as anti-death beings. Think about how the magnificent elk was honed by the presence of the wolf. Perhaps what the wisdom traditions point to is the psychological equivalent of this biological response to death threats from other organisms. This psychological parallel happens all in one brain. Within a brain the idea of death can bring forth not just fear but a response to fear. Is this response what the traditions view as the gain from contemplating death?

We have seen evidence that consciousness involves special parts or networks of the brain. It is a flickering dynamic core in which certain brain organs are crucially important. But what is the function of consciousness? If a function exists, does it help us understand what benefit could derive from contemplating death?

The function of consciousness is not easily figured. For instance, one might simply say the function is so we exist as selves. But such existence might be a result of consciousness, not its primary function. In the technical literature on consciousness one even hears some speculation that consciousness itself is what is called an epiphenomenon. Just as a lightbulb's function is to produce light—with heat as a byproduct, its epiphenomenon—so the torrent of activity in the cognitive unconscious from moment to moment might throw off consciousness.

But most neurobiologists believe that consciousness does have a function. Something so important would not have come about in evolution were it not vital to the whole brain and our biological beings.

One cognitive scientist who has made a case for the function of consciousness is Bernard Baars, who elucidated various key parts of the brain that produce it. He likens consciousness to a spotlight on a theater stage. Players enter the spotlight, then disappear as new players come in. Where do the players who are so integrated in the spotlight come from? The sights, sounds, thoughts, imagery, and more enter the spotlight from the offstage darkness, from the unconscious. Several players can perform together in the spotlight, which is the synchronization of parts of consciousness, such as sights and sounds. If the spotlight is too empty, it is common for more players to shove their way in. (Or perhaps they are pushed by the stagehands of unconscious emotions and urges whose actual hands can be glimpsed sticking into the spotlight as they shove the players in.)

What happens to the activities of the players in the spotlight of consciousness? Do their words and actions just vanish with no effects? Taken alone, the spotlight is a passive metaphor. It is a metaphoric field of light, perhaps cast by the thalamic intralaminar nuclei (the light is not in these nuclei, but is created by them via their connections to the rest of the brain) and existing as Edelman and Tononi's flickering dynamic core. But even if we think of the spotlight not as a site but as a dynamic geometry from the physiology of the brain, it all still seems so passive—too passive. What happens to the actions of the events in consciousness?

Baars says that the spotlight possesses an active function with a profound biological purpose. And here we come to the feature that will be key in a suggestion about how to think about death in relation to the self. In a nutshell, Baars says, consciousness "is the publicity organ in the society of mind."[36]

Consciousness as a spotlight, a lit-up gathering of players, allows a vast audience to all witness the same material. More specifically, consciousness as a spotlight "serves to disseminate a small amount of information to a vast audience in the brain."

What audience? Why, the audience sitting there in the dark, in other words, in the unconscious. The audience consists of many parts of the unconscious that we never see, such as memory systems, a variety of interpreters, processors of automatic behaviors, and motivational complexes. Furthermore, it is obvious that members of the audience are not passive. Because the brain's circuits are cycles, it seems apparent to me that some members of the audience can run backstage to talk to the various directors, assistant directors, set designers, electricians, and stagehands, and thereby influence these offstage determiners of which players should leave the spotlight and which players should enter, and the player's relationships and contexts when they are on stage.

I find a similar view in the ideas about consciousness from Edelman and Tononi: "When we become aware of something, whether it is an uneasiness in how we balance ourselves as we walk, a flutter in our stomach, a mistake in our reasoning, or the slow emergence of the pattern of an object out of a random-dot stereogram, we can make use of that information in a large number of possible ways that can trigger all kinds of behavioral responses. It is as if, suddenly, many different parts of our brain were privy to information that was previously confined to some specialized subsystem."[37] Via consciousness, information in some specialized subsystem (one of the players) moves onstage into the spotlight of the dynamic core and then becomes available to many different parts of the brain, in other words, to the audience who all witness what is in the spotlight.

Following these ideas, we can say that our total being requires consciousness. In fact, our cognitive unconscious requires consciousness to operate, because the cognitive unconscious, consisting of both players offstage and the audience, requires the spotlight as both a concentrator and a disperser of information, to coordinate the whole brain. The anatomy and dynamics of exactly how this occurs, and to what other species it extends, is not known at this time. Some say, for example, that a key behavior of the brain is the ability to make synchronized rhythms, which bind different parts of the brain together as items become conscious.[38] Such rhythms might be the reason Baars can use the metaphor of the audience, all witnessing the same actor at the same time, thus gaining coordination. Perhaps consciousness is a particular form of synchronized convergence that functions to distribute its patterns to a multitude of other parts of the brain. Moment to moment, we live as loops between the unconscious and conscious.

Imagine yourself sitting at dinner in conversation with friends. You are listening to a story. Suddenly your own similar, but you think better, story pops into your consciousness. It must have come from some pattern-seeking part of your unconscious. OK, thank you. Now you have an urge to tell your story. The urge is also from the unconscious, because it just wells up and then is felt, you didn't first consciously decide to feel it. You might want to push right in as some conversationalists do but hold yourself back, and start looking for the polite moment to begin. How does this search work? You might think it is all conscious, but much is unconscious. This is not the first time in your life that you have silently said "Wait" to one of your own stories just bursting at the seams of your mouth to get out. You have had practice at waiting for the discreet moment. There are habits for such a task in your unconscious, just like the habits for riding a bicycle. The impulse to speak from the unconscious keeps pressure on the consciousness, which flows out to the inner audience members of the unconscious, who, skilled with the necessary habits, are responding and gearing into action as they have many other times in similar situations in your life. Hearing is not just listening to the meaning of the words of the outside speaker but also to the cadences and pauses, and predicting when the end is near. Your eyes are watching others out there, and consciousness distributes certain aspects of their behaviors to many parts of the brain, where analyzers in the unconscious are making rapid notes on who else might be twitching in their seat, and judging whether these twitches are from hemorrhoids or from someone else fighting back the urge to speak. If the latter, then better watch them more carefully.

Thus our own self-contained little conversation, with speaker and listener all in the box of the skull, might have a purpose. It often seems to—we feel we "get somewhere" when we, say, think through a task. Consciousness is vital for this journey. It is not that no parts of the unconscious are connected without consciousness, but that consciousness is a function that provides connections otherwise absent, important connections for being human.[39]   So what we put into consciousness is of utmost importance.

If we think about death, does the unconscious respond with fearful actors coming from offstage? Or can the brain learn a different kind of response? It can learn that fear doesn't help in our human situation because there's no fight or flight that will remove the skeleton from one's life at the end of the path. But can the idea of death produce an emotion or cognition that works wonders for our being?

During my wintry walk at the time of my poisoning by carbon monoxide, I realized truly that someday I would die, really die. Then a feeling gushed forth in me in response, like a spring of water from the earth, that I would rather live now for another day than already be dead. Well, this was one of those simple statements such as the fact of the inevitability of death itself that can be intellectually known as true or felt to the deepest bone as TRUE. In the midst of despair that all I loved to do would be thwarted, maybe soon, I fought back horror with the recognition, the joyous recognition, of how happy I was to have been alive, to have been granted the privilege from this sublime universe to have life, to have consciousness, to have selfhood, to have known others. I was thankful, so very thankful. And would continue to be thankful if granted another day. I accepted that death might be the penalty to pay for the advantages of all other human powers, not just to see and to think, but to be alive in the most raw, undefined sense of the word, to be conscious, to be me. If death was the fee for having life to be paid at the end, then I accepted the bargain with open arms.

So I propose that we greet not Death but Life with simple gratitude. This is a gratitude that can flood and support and fill one's being as the sun does the sky, as a tree supports a flock of migrating birds, as air bathes our lungs. This is my answer to what I have constructed as my monist belief system, based on what I know about the brain, about death, about the self. Since that evening, I have experimented with it countless times and the results have confirmed my discovery. I let gratitude well up in moments of despondency about the transitory nature of life. Gratitude is the response I give for the fears engendered by death. I have found that one cannot overdo gratitude. Gratitude allows me to face death more enlightened.

I spoke with Michael Lewis, a distinguished professor at Rutgers University and an expert on the so-called self-conscious emotions, or what could also be called the social emotions.[40] Self-conscious emotions occur later in a child's development than the primary emotions. Debates surround the makeup of the primary list, but fear is inevitably there. So too, typically, are anger, sadness, joy, surprise, and disgust. These primary emotions will be expressed by the infant between birth and six months of age. Thus they are biological givens independent of social training. In contrast, the self-conscious emotions come about later, around two or three years of age. They include pride, shame, guilt, embarrassment, envy, and empathy. These depend on reactions to others, on the opinions toward oneself from others, and on how one sees oneself in a social world of others.

I asked Professor Lewis about gratitude. It was not on his list. He said it was difficult to define. Is it an emotion or a cognition, for example, he mused? He settled on calling it an "emotional term." However, he was definite that this emotional term refers to something social. Gratitude is not primary like fear. It is more developmentally advanced. It requires interactions with others for children to learn it. (I can hear one parent say to the young child, "Now say thank you to your brother for sharing the ice cream. I said, say thank you!")

Lewis also puts gratitude on the side of what he calls "life forces." Life forces are emotions or emotional attributes that promote life in a physiological manner by reducing stress. They include optimism, happiness, and interpersonal connections, as well as gratitude. The second side of his two great categories of emotions or emotional attributes he calls "death forces." These include anger, depression, shame.

It seems to me the trick in contemplating death is to turn the awareness of it into a response that is a life force. Gratitude, based on my experience, is an excellent candidate for this transformation.

Gratitude has been considered from the viewpoint of evolutionary psychology. This new science studies the deep roots of people's behaviors as innate, genetically based properties evolved over perhaps a couple of million years. Specifically, the intense level of social interaction whereby human ancestors helped each other has been called "reciprocal altruism." To social theorist Robert Trivers, gratitude evolved as one of the "mechanisms to regulate reciprocal altruism." Specifically, according to Trivers, "gratitude has been selected to regulate human responses to altruistic acts, and that emotion is sensitive to the cost/benefit ratio of such acts."[41]

According to Trivers (and picked up by M.I.T. professor Steven Pinker), we are evolved to feel gratitude when someone gives us something that is either a great benefit to ourselves or a great cost to them. Pinker says that gratitude "calibrates the desire to reciprocate."[42] If you feel that someone carries gratitude based on your gift or action, you know they are more likely to reciprocate. And you are more likely to give again. Feeling gratitude might train the brain to reciprocate, in one of those publicity launches from consciousness out to various parts of the unconscious that coordinate the brain and body.

What I find interesting about gratitude in the context of thinking about death is the fact that the gratitude I feel toward life in general is not aimed at any person. Life itself—the biosphere, for instance—has given me a gift. Gratitude can be toward the universe. So if our capacity for gratitude evolved deep in the human past as a feeling associated with human relationships, it has now been extended to a wider world, in a sense the abstract world of all of nature.

Lewis thought this extension could be easily explained in the case of gratitude to a god, for those who cherish such relationship. In traditions where God is a "superperson," he noted, then feeling gratitude toward God is still a social relationship, thus keeping the originally evolved human capacity for this social emotion.

Lewis also recalls Job, God's perfect servant whom he tortured and whose life he ruined in seemingly senseless ways. Lewis says the message is that one should be grateful for the social contract with God, not for what God does. Seems a terrible bargain, until you consider that God chose you. Similarly, we rejoice in our contract with life, death our half of the deal.

Gratitude, clearly, is felt when what is received is of great value to the receiver. In the case of the atmosphere, gratitude comes forth even though the gift does not cost the receiver anything. We were just born on this planet. But one can consider that all past organisms who lived and died gave chemical structure to the atmosphere we have. Without the atmosphere we could not live. Therefore to all that comes for free from nature, even though no reciprocal price is obvious, we are grateful because of the irreplaceable benefits we derive during life.

Oh, does life become sweet. Feeling gratitude makes life sweeter and sweeter and is a way of making friends with the idea of death. Gratitude is important to cultivate. It is a key player to be kept in the spotlight of consciousness, or at least waiting in the wings just offstage, ready to come into the spotlight at a moment's notice, during moments of despair or fear when the idea of death itself comes forth. Gratitude is not just an antidote to the concept of death, but an emotion that can be intensified throughout life as death is kept conscious. Gratitude can mature. Because it is a social emotion, one feels social with anything one feels gratitude toward.

Deferring once again to religious traditions for experimental examples, as solutions to the problem of mortality, it is pleasing to discover gratitude as one of their main themes for the way to live. Universally, people give prayers of thanks. They thank variously named gods for what has been given. Thanks might be offered for external bounty, perhaps for green harvest or the constant sun. Gratitude can be directed internally, as thanks for thoughts and feelings. In the religious traditions, hardship, too, can be praised for the lessons it teaches. When bounty and suffering are both praised, then what is being thanked is nothing less than all of existence itself. Everywhere, in hymns of devotion, gratitude is encouraged and even insisted upon. We should use these traditions as guides. Sure, we might eventually find it illuminating to put those folks who are full of gratitude through brain scans. But in the meantime, let us give thanks for what we already know.

Cultivating gratitude in consciousness will broadcast that feeling to various regions of the unconscious. Gratitude is thus used as an organizing principle for our diverse psychological machinery. For dualists a brief thanks to God before a meal, or for monists a thanks to nature when standing at a waterfall, can shift the moment into a more memorable state, preparing the brain for a continuously and ever-renewing, vivid, unified, all-senses-turned-on experience.

Gratitude can foster other emotional attitudes and values. For example, religious traditions suggest many crucial attitudes in addition to gratitude. These include love, charity, honesty, right actions, right speech, family piety, compassion, humility. However, I have noticed that these can be more easily brought into one's behavior when gratitude sets the underlying emotional tone. These other attitudes might be called the children of gratitude. With gratitude one is more likely to be kind, express love, take responsible action, and as well be astutely conscious of cultivating these precious off-spring. In any case, even if not necessary for their births, gratitude can help to reinforce these other desirable attributes.

For those of the dualist persuasion, gratitude is welcomed not only for its earthly benefits but because it can induce a state of thanks directed to the higher powers on the other plane of the afterlife. Thereby one is directed toward proper acknowledgment of those powers, should they exist, and a person full of gratitude and the other graces will presumably enter the afterlife with more virtue.

But gratitude is especially important to monists, like me, who are in more dire need in this life for a way to counteract the fear of death as final annihilation. In the case of monists, gratitude might be vital for well-being and even sanity in the face of mortality. As gratitude spreads from consciousness into the unconscious, and as it allays the fear of death, it changes the totality of the self to a state of desiring even more gratitude. We are definitely evolved to use what works for peace of mind and psychological development. Thus to the extent gratitude creates a pleasurable state of being that counters the anxiety about mortality, it will be called upon more often.

I want to be clear that gratitude is not simply a reaction to fear, like putting up a shield against an oncoming spear. Gratitude is more like grabbing the spear as it flies toward one's body, and then using the momentum to turn around and around with the spear in hand and dance oneself into ecstasy. The spear is not knocked aside to the ground, it is played with.

Others have noted the importance of gratitude, either as a valued response specifically honed when confronting the finality of death, or as an emotional stance toward existence and the munificence of nature.

I may not believe in life after death, but what a gift it is to be alive now." [43]

That is science writer Natalie Angier, speaking from a view that she calls "transcendent atheism." Here is cell biologist Ursula Goodenough, as she contemplates mortality and the fear it engenders:

> *As a religious naturalist I say, "What Is, Is" with the same bowing of the head, the same bending of the knee. Which then allows me to say "Blessed Be to What Is" with thanksgiving.* [44]

Philosopher of evolution Loyal Rue, also dealing with death, offers the following:

> *. . . I will submerge the absurdity of death in gratitude for the wonder and wisdom of life.* [45]

Finally, the words of Zen Buddhist scholar Daisetz Suzuki and Greek writer George Seferis, as they consider bounty received from nature:

> *The sun is the great benefactor of all life on earth, and it is only proper for us human beings to approach a benefactor of any kind, animate or inanimate, with a deep feeling of gratitude and appreciation.* [46]

The day before yesterday under my northern window I saw two shy little blossoms on an almond tree—the first almond tree in blossom. On the mountain the cyclamen turn to leaves; in the sea, the dark blue fluttering of the kingfisher. I give thanks.[47]

PART TWO

culture

NOBODY
JUST
DIES

A number of archeological sites have been found near where I am writing in New Mexico. One sits on a low bluff of private land, overlooking a river. It was excavated over a decade of summers by an amateur archeologist, who uncovered the remains of what had been an impressive adobe complex of thirty rooms. This glimpse into lives of those about a thousand years ago who hunted, farmed, loved, and sang was made particularly poignant by the numerous burials discovered.[48]

The burials are located in what would have been the earthen floors of most rooms, which in many cases were used for living, as evidenced by the remnants of cooking hearths near the rooms' centers. Burial in the floor was common practice for the Mimbres culture, who flourished in this part of the Southwest. They are famous for their painted black-on-white pottery. During their "classic" era, the Mimbres depicted animals so realistically that very often the species of bird, fish, or mammal can be identified.

Most burials have ceramic bowls associated with them, one of which especially intrigued me when a neighbor let me look through photos. It had an exquisite black caterpillar painted on the inside bottom where the last dram of corn soup would have rested. More striking, though, the bowl was broken. It had an irregular hole about a half inch across, right at the tail end of the caterpillar. Cracks radiate away from the hole out into the base of the bowl. Is this unfortunate? It turns out that such a break is the signature style of Mimbres burials. These broken bowls are what the archeologists call "killed." To kill a bowl for burial, a hole is deliberately punched into its bottom. Most commonly a killed bowl would be placed over the head of the corpse like a ceramic skullcap.

The caterpillar bowl was found with the burial of a man estimated to have been about my age, a real elder back then. His original resting place had been greatly disturbed by the subsequent placement of an abutting corpse. Altogether there were thirteen in that room, situated mainly along the walls, with the majority clustered in the southeast corner. Out of the total, only five were adults. So much decay had taken place and so few bones remained intact that, except for the man with the caterpillar bowl, sex was impossible to determine. The other burials included a child of about ten, a child around one, an infant of six months or so, and five fetuses.

Cultures that dealt with their dead in a manner that imperiled the health of the living would not have thrived, and burial is one of the time-tested methods for maintaining hygiene. What is hygienic can vary, though. Nigel Barley in his book *Grave Matters,* which surveys the astonishing (to me) customs of death around the world, reports on his experience with a

Torajan granny who had been "sleeping" for three years in a home in Sulawesi.[49] The custom of wrapping the dead in "vast amounts of absorbent cloth to soak up the juices of putrefaction," while resources and money are being collected for the next stage of the funeral, sometimes leads to an extra piece of furniture for awhile. Barley was requested to greet the granny in what "looked like a bundle of old clothes," which was being used as a shelf for cassette tapes.

For the Mimbres, burial sites outside the dwellings might have been more sanitary. But the region is rocky, and digging a grave deep enough to frustrate the claws of hungry scavengers would have been difficult. In that case, why bury at all? What would be wrong with letting wild animals take the flesh (and bones with their good fatty marrow)? We can assume that such disposal, though a form of excellent hygiene, might have upset the people. Thus the fact that the Mimbres chose in-house inhumation makes us suspect that hygiene by itself was not the only goal. Indeed, with the Torajan granny, hygiene was clearly not the only issue.

With killed bowls and other funeral customs, we see culture mightily exerting its needs. Why bowls? The Mimbres people could have placed a deer antler or spearhead in the grave. But no. Their punched-out bowls meant something special. Exactly what? Nobody knows. The Mimbres had no writing. But Nigel Barley has found that smashing pottery at death ceremonies is quite common around the world.[50] It can be used to prevent the dead from returning; it can be a ritualized metaphor for the finality of death itself. Specifically with regard to the Mimbres, archeologists who have studied death ceremonies of contemporary Native American tribes in the Southwest, such as those of the Hopi and Zuni, surmise that the purpose of the Mimbres kill hole was to release the spirit of the bowl at the same time as that of the deceased. Thus the proper journey of the person's spirit to the next world was ensured.[51]

The grave goods of the Mimbres were relatively simple. In contrast, consider the complexity of the spectacular grave objects and funeral preparations of ancient Egypt. Grave sites were filled with wonders that today bring throngs to museum shows. The corpses themselves were multiply swaddled by the imprint of cultural needs and symbols turned into matter. Specifically, the dead bodies were mummified, wrapped, and set inside inner coffins, which in turn were set within elaborately painted outer coffins. Then these assemblies of layers were placed inside ponderous stone sarcophagi carved with delicate images of Egyptian mythology.

These sarcophagi were often set underground or within cliffside stone tombs. Murals on tomb walls can show musicians that play for the deceased in the afterlife, food to be eaten, boats for the journey into the next world, and the weighing of the heart and its judgment by jackal-headed god Anubis. Often the tombs contained actual food and boats. Above ground, the tombs could be as large as the giant pyramids themselves. The biggest of the three at Gizah held the corpse of King Cheops, along with the requisite, amazing riches, in a chamber near the center. The symbolism of the pyramids themselves refers, in part, to the mountain, which joins earth and sky—the changeable and the eternal, hence mortal and immortal. Another meaning relates to the rays of the all-powerful sun, more evident in the pyramids' pristine days when their sloped sides were jacketed with glistening limestone and their peaks golden. Thus solar rays spreading down to earth were concretized as the giant stone prism around the dead Pharaoh, who rested within the rays' embrace and who was cared for during death's aftermath with the help of rites conducted in adjacent mortuary temples.[52]

What we see in these cases, from materially simple to materially complex, is death bringing life to culture. In both the stone pyramid and the ceramic skullcap I see that nobody just dies without being swaddled in one final ennoblement. And the key part of this wrapping is the psychological aspect, the private yet socially coordinated experiences and emotions of the people who surround the deceased.

We might look at the effect of death as almost like that of a small particle of rock being dropped into a supersaturated solution of some mineral ion. The particle acts as a seed to start a crystal rapidly growing. In a society, consider the people out and about doing their business, like ions of dissolved minerals in solution. They are supersaturated in that all exist in a state of potential upset, due to their recognition that the death of a friend or relative, as well as their own, could occur at any moment. Suddenly an actual death plops into reality. And immediately the ions crystallize around the dead person in organized rituals. Each ritual is true to its culture, as crystals of calcium or silicon will be true to their form, given their own specific chemical characteristics such as bonding geometry. Social life thereby forms around individual death.

One of the twentieth century's foremost anthropologists and scholars of what used to be called "primitive" religion promulgated a close tie between death and religion. In fact, Bronislaw Malinowski saw death as more than a particle that catalyzes the growth of rituals. He saw the existence of death as the actual, greatest overall source of religion. This is "death, thus social life" on a grand scale. Some of the impetus to create religious social structures, such as funeral rites, derives from the fact that so many religions promise immortality, a scenario of "death, thus afterlife." Malinowski:

The testimony of the senses, the gruesome decomposition of the
corpse, the visible disappearance of the personality—certain appar-
ently instinctive suggestions of fear and horror seem to threaten man
at all stages of culture with some idea of annihilation, with some
hidden fears and forebodings. And here into this play of emotional
forces, into this supreme dilemma of life and final death, religion
steps in selecting the positive creed, the comforting view, the cultur-
ally valuable belief in immortality, in the spirit independent of the
body, and in the continuance of life after death. [53]

But what if the afterlife does not exist? Or what if one does not
believe in the afterlife? Religion of course would be much changed. But
what about funeral rituals themselves?

The rituals and social structuring around death would continue, in
my opinion, given that Malinowski also notes that funerals serve purposes
other than to help imprint an afterlife belief system. These purposes create
a social life as reaction to death in ways not tied to immortality. Specifically,
two somewhat opposing emotions come into play.

The first is a "shattering fear of the gruesome thing that has been
left over." Obviously this can feed into a belief in the afterlife. But fear also
engenders social mechanisms to bring the horror of death under control in
an immediately practical sense. This is because the horror has the capacity
to cause individuals to run away from their tribal village, or perhaps to
viciously destroy the corpse which so threatens their life-preserving
instincts. According to Malinowski, "if primitive man yielded always to the
disintegrating impulses of his reaction to death, the continuity of tradition
and the existence of material civilization would be made impossible."[54]

Thus the funeral rituals create order and combat potential chaos. They are highly ritualized and crucial for society. Indeed a recent Peace Corps volunteer who returned from the Ivory Coast of Africa and had witnessed funerals of different tribes told me that of all the rituals she saw, the funerals were the most important to attend and the most important for the cohesion of the group, far surpassing in significance birth rites or marriage ceremonies.

Coalescing people into a dynamic social crystal around the corpse forces them to face death. And thus the moment functions not just to calm their fears but to awaken new aspects of each person as well. Here the social life created by death becomes, for each individual, a bit more transcendent. Death serves to awaken the consciousness of the living. It's the lesson that Buddha saw from his chariot ride. It's the lesson that Jesus taught in the parable of the rich farmer. As Greek writer Neni Panourgiá put it,

> *Anyone else's funeral could be your funeral. In any one of the caskets you see being lowered into the ground could be your body. And it will be. Whether you speak of it or not, whether you name it or not, the devastation of death is counterbalanced only by the totality of its existence. Naming it, speaking it, is simply the act of memory and the means by which you measure your relationship to the past.* [55]

Panourgiá speaks of memory. The memory includes one's autobiography and, crucially, memory of the deceased. And here we find the second of Malinowski's emotional elements coming into play. Very different from the first but complementary to it, this aspect is the "passionate attachment to the personality still lingering about the body," literally the love still felt for the deceased. This feeling leads the survivors to treat the corpse well, to caress it, dress it in fine, clean clothes, to treat it lovingly not just to help its way in the afterlife but for its own sake and theirs, to continue the tender emotions.[56]

The tie is broken, good-byes are said. Yet the tie is also maintained. Memory is cemented. The dead live on as memory within survivors, and the social life born in the immediate aftermath of death helps consolidate the memory in each individual via a collective experience.

For example, in the world of the ancient Greeks, the major (if not total) form of immortality was in the remembrance of the dead by the living. We still have examples of tall stone shafts (stelae), erected over two thousand years ago, with carved words and relief figures. These stelae were especially important for dead youths, because they had no children to remember them.[57] Thus the stelae with words and images served to broadcast their memory.

Today, in memorial services, the function that helps provide continuity in emotional attachment to the deceased unfolds when we find friends and relatives rising up to speak of personal memories in words that flow outward to the gathering. Because death rituals throughout the world are collective, shared grief at a shared loss, this function operates whether or not a people believe in an afterlife.

The manner in which such memories are institutionalized affects the living more than just by way of specific recollections of the dead. The living see not only what is remembered, they witness the tone of the funeral. Worldwide, the most universal expressions are of grief and respect. When the living see that others lament the dead, they are consoled about their own future deaths. Why should anyone, or at least those who do not believe in an afterlife, care how they are remembered? Well, I do. The concern seems almost instinctual. Indeed, the prospect of one's own death that comes up in daily life via ordinary human awareness is frightening enough without seeing someone dead. But when such trauma occurs, the fear is elevated and then assuaged by respectful burial practices. We learn gratitude for our relationships with others from such ceremonies and anticipate that others will similarly feel gratitude toward us when we die.

I am reminded of a joke. A comedian is bemused by the fact that we are taught to speak only good of the dead. "What's that you say? He's dead? Well, good!"

What makes this funny is that we probably want to say it more often than we do. Very improper, to say the least. It is also funny because it violates basic instincts and puts us in touch with our fears. It would be unnerving to live in a society whose norm was such an attitude toward the dead.

That the dead cannot be treated with the disdain of our comic is quite obvious, you might say. True. But only because such treatment is abhorrent and unimaginable. Such an entrenched feeling reveals something profound. It shows how connected we are, how dependent we are on existence in the thoughts of others. Death can explosively bring to the mind's surface such dependency. Somewhere in the human mind, not as cool calculation but as boiling emotion, is a golden rule of death: Do unto dead others as you would like to have done unto dead you.

These common human reactions—fear of and love for the corpse—determine the complex ways in which, globally, we all do something with the dead. Nobody just dies. Because society organizes itself around death, we can say that death creates life on a social scale. In the aftermath triggered by an individual's death, social structure flowers forth. Turning to Malinowski again, "The ritual despair, the obsequies, the acts of mourning, express the emotion of the bereaved and the loss of the whole group. They endorse and duplicate the natural feelings of the survivors; they create a social event out of a natural act."[58] Think about it. The most private act of an individual—death—is turned into a public event.

In funeral rites we see how a controlled act of social life surrounds an uncontrolled act of death. One of the modern tribes of Native Americans, examined for insights into meaning of the Mimbres burial practices, offers a glimpse into an act of social life built around a death that is itself controlled. This involves animal sacrifice. In the case of the Hopi people and eagles from around Wupatki National Monument in Arizona,

> For generations, young men have scaled cliffs each spring to gather eaglets, which are considered messengers between the physical and spiritual worlds. The eaglets are reared until July, when they are sacrificed to send them to their spirit home.[59]

In a Hopi village recently I saw an eagle and a redtail hawk together in a cage on the roof of a house. I was told that the birds, in general, are indeed captured young and raised for ritual sacrifice, in which they are killed without blood spill. Their feathers are used for special attire and objects in ritual dances and other practices. Because it is a lot of work to raise the birds (they are fed rabbits from hunts, for example), all who help provide food are granted a share of the magnificent feathers.

Funerals and sacrifices are similar to each other in that both involve social gatherings and rituals—thus acts of social life—around death. True, the Hopi case is a rarity today, but though ritual animal sacrifice has decreased in importance, it was of major import to much of the world's population in the past. Just how powerful such rituals were the world over during much of human history was brought home to me when in modern Rome I visited a monument from ancient times.

It was the age of Augustus. The noble leader himself, having returned from a bloody mission of pacifying border regions, was acknowledged by the Roman Senate to have created a general peace, eventually known as the Pax Romana. The Senate decreed that a marble building be constructed to honor this peace yearly. It was called the ara della Pace Augusta, the altar of the Augustan Peace. Today it is simply called the Ara Pacis, the Altar of Peace.

The outer walls of the white, rectangular temple bear relief carvings, the two long walls featuring processions of robed Romans. Mythic scenes flank doorways on both shorter walls. The walk-around altar inside the Ara Pacis can be reached from either doorway. Up the steps the panel to the right of the main entry depicts the Trojan hero Aeneas. It captures the glorious moment when he set foot back in his homeland of what is now Italy, following the war saga of the *Iliad*. In thanks, he is about to sacrifice a sow. The scene would have been visible to the procession of the Pontifex Maximus and the Vestal Virgins as they climbed the short steps and passed through the doorway for the annual ceremony.[60]

Inside, high along the walls, arrays of garlands are strung between series of bull skulls, all exquisitely carved in marble. Through the "back" door another group would have entered. This entry had a ramp to it because some designated for that door would have had difficulty climbing steps. As shown in a relief of the ceremony, the sacrificial party entering through this back entrance included priestly figures and killers with weapons—mostly knives but also a club. They pulled and pushed the victims toward the altar inside: a sheep, a bull, and a heifer. The group split to right and left, and then waited along the inner corridors during purification and holy recitations. Then the animals were killed. The inaugural bloodflow at the Altar of Peace was in 9 B.C.E.

What is going on? Animal death is cultural life? This is what I wondered as my eyes wandered across the art of the Ara Pacis, with the sculpted cattle skulls, the mythic scene of Aeneas and the doomed pig, and, via imagination, the sacrificial procession itself that took place long ago.

Classicist Walter Burkert, referring to the Ara Pacis, showed palpable awe at the fact that "the most refined Augustan art provides a framework for the bloody sacrifices at the center." Furthermore, noting the wide occurrence of animal sacrifices, "it is astounding, details aside, to observe the similarity of action and experience from Athens to Jerusalem and on to Babylon." He could have included China, India, Africa, and many other places around the world. For example, from more than two millennia before Augustan Rome, on the small Mediterranean island of Malta between Italy and Africa, the stone temple complex called Ggantija contains evidence of a cult of animal sacrifice. Beneath the threshold slab of the complex were found the horns of a sacrificed ox and an offering bowl.[61]

The Ara Pacis was not built to kill animals. It was built to com-memorate the peace of Augustus during a celebration accompanied by ani-mal sacrifice. It's just that in those days celebrations and rites, from family to state, often involved animal sacrifice. Sacrifices were both private and public.

East of Rome and earlier than the days of the Ara Pacis, people of the rugged archipelago of ancient Greece also engaged in sacrifice perva-sively. Scholars of the records of Athens of the fourth century B.C.E., for example, have concluded that the typical citizen rarely ate meat except at public sacrifices.[62] The estimates suggest that oxen might have been sacri-ficed sixteen times over the ritual year, averaging out to about 1 ox annu-ally for every 25 adult males. Other public sacrifices used goats or sheep. Private rituals could take place with smaller animals as well, with meat dis-tributed to friends who participated in small, solemn events. In fact, although some less desirable, leftover, parts of the animals would be sold to shops for the open market, there were virtually no prime cuts consumed apart from the sacrificial system.

According to Burkert, the author of *Homo Necans,* a book about ancient Greek sacrificial rituals, a typical sacrifice to the Olympian gods can be pieced together fairly well from numerous descriptions.[63] It began with a procession. The participants moved as a single body, in rhythm with singing. They conducted the gaily decorated animal to the sacrificial stone, the "altar." In this atmosphere of incense and flute music, a virgin girl led the way, carrying a basket.

At the site, the participants circled and passed round a basket and water jug, from which they washed their hands. The animal was sprinkled, which usually induced it to shake its head; by this willing nod it agreed to be sacrificed. From the basket the participants took barley grains. After a silence, then spoken prayer, they flung the grains over animal and altar.

In the basket, beneath the grains, lay the knife. The animal would often be enticed toward the barley, which was taken to be an act of aggression toward the hidden knife. Thus the knife would have to defend itself.[64] First, it was used to deftly slice off a few hairs from the beast's forehead, which were then thrown into the fire burning atop the altar. Then the women all cried forth, with a wailing scream, which drowned out the animal's own cry as the deathly, climactic cut across its throat was made.

A small animal such as a chicken might be lifted over the altar. For large animals, such as sheep or ox, the blood was caught in a bowl for its sacred dispersal. The flowing blood was not permitted to touch ordinary ground, but only the altar, the fire, or the sacrificial pit.

The animal was carved up. Rules specified the fate of each piece. First, the heart was placed on the altar, sometimes still beating. A seer would interpret the liver. Most of the animal was cooked, after careful positioning on the altar, and much was eaten on the spot in a communal meal, internal organs included, except for the inedible bile. After more rituals involving wine, cakes of bread, and proper placements of the bones, the sacred event wound down and participants dispersed. Earnings from the sale of the skin would go to the sanctuary.

Animal sacrifices were also vital to the worldview of the Hebrews, at the fringes of the Pax Romana on the eastern side of the Mediterranean. By the time of Augustus, the Hebrew sacrificial rituals were restricted to the governing temple in Jerusalem. But these blood rites of theirs had a long history, and were of diverse types. For example, there were sin offerings, trespass offerings, meat offerings, and peace offerings. The events almost invariably involved a kill. Animals that could be used were bulls, male or female or kid goats, doves or young pigeons, male or female sheep (and without blemish, a condition for all, if you please). In one type of sacrifice, called the burnt offering, none of the meat whatsoever was consumed by the giver or the priest. All was burned.

These Hebrew rituals were first spelled out as instructions from the Lord, in Leviticus, the third book of Moses. Direct words from this book give a powerful sense of the event. Here is one example, a burnt offering:

> *And Moses brought the ram for the burnt offering: and Aaron and his sons laid their hands upon the head of the ram. And Moses killed it; and he sprinkled the blood upon the altar round about. And he cut the ram into pieces; and Moses burnt the head, and the pieces, and the fat. And he washed the innards and the legs in water; and Moses burnt the whole ram upon the altar: it was a burnt sacrifice for a sweet savour, and an offering made by fire unto the Lord; as the Lord commanded Moses.[65]*

The Hebrew burnt offering was a true sacrifice in the modern use of the word, meaning voluntary loss, giving something up, because the entire animal was consumed not by people but by flames. Only God got the meat. The function was symbolic, in other words, psychological. But except for this special case of the burnt offering, in Hebrew sacrifices some portion of the animal was returned to the offerer or kept by the priests for their consumption. However, there was a giving up because the sinner did not necessarily plan on butchering an animal at that time (or donating any of it to the priests).

In the ancient Greek sacrifices, the gods received only the cooked gallbladder, fat, and bones. To contemporary comedy writers of Greek theater this pittance seemed a blatantly raw deal for the gods. Thus, the sacrifice was clearly at least a partially dressed-up excuse for a communal meal of meat. Perhaps even the monthly lamb during the calendar cycle of Confucian times in China could simply be considered as a wrap of culture around the real function of satisfying the lust for a barbecue.

It is not known what happened to the meat at the Ara Pacis. But the building was too expensive to be principally a slaughterhouse once a year. All this marble just to be bloodied by a butchering? No. Recall that the killings in the Ara Pacis were part of the annual commemoration of the start of the great peace. Thus in the context of refined marble art, the sacrifice served the Roman psyche.

Yet the fact remains that the sacrificed animals would have been food as well. We should consider the killing and eating as a ritualized pairing of actions. What are the roots of this pairing? Originally, according to Burkert, "the necessary combination of death and eating appeared only in the hunt."[66]

In the days of deep prehistory, celebrations following a successful hunt would likely have preceded, in cultural evolution, the primitive funeral. (With regard to hunting ceremonies, we are probably talking hundreds of thousands of years prior to agriculture and its own attendant celebrations for successful harvests. The first burials were about a hundred thousand years ago.) From the dawn of human culture, a big kill would have sparked the tribe into general joy. The animal would have been carved, exposing blood and guts and muscle. Its consumption was a time to stuff human bellies with as much fresh or freshly cooked meat as possible. A feast ensued. Perhaps afterwards, in the air of unusually heightened satisfaction, the hunt would have been reenacted using theater-type skills. Thus we could have witnessed singing, dancing, and miming of the drama of the stalk and kill. Some child would shout out and sign with her hands, "Again, again!"

From the hunt how did the sacrifice come about? The Hebrew ram, the Greek ox, the Roman pig, the Chinese lamb—these were not wild animals hunted under circumstances in which their death was uncertain. These were domesticated animals. With animal husbandry, a successful "hunt" could be accomplished at any time desired. A profound shift in the human relationship to animal death had taken place. But it was still death, with blood and guts. It was still somewhat a cause for celebration, because meat was put on the table. From this pragmatic fact it is not so huge a step for an ancient culture to think that such a kill could benefit or even feed gods. It is collective gratitude. And perhaps the kill, which satisfies the human stomach and thus ultimately seems so good, could also bring about a desired outcome such as a military victory, or could atone for sins against god or community, or could serve as thanks to Invisible Controllers for their gifts of past boons.

In sacrifice, the moment of death can be precisely designed. The subsequent flow of blood shocks people's consciousness; it focuses their minds and makes them feel almighty because they have just participated in an instance of control over death. Perhaps the certainty of the sacrificial kill helps people overcome anxiety about the ultimate uncertainty of their own eventual deaths. In jolting people's webs of concepts and feelings involving death, the sacrifice is like a funeral. For one, both often involve processions. For another, consider the prayers and music. Also, in both food is important. As Burkert has noted, the "most widespread element in funerals—so obvious it may seem hardly worth mentioning—is the role played by eating, i.e., the funerary meal."[67]

Human sacrifice is about pure meaning, not food, except for cannibalistic rites. In the history of human sacrifice, biological death nurtures the life of culture symbolically.

There is mention of human sacrifices in the Bible. For example, during times of evil abominations, the Hebrews made their sons or daughters "pass through the fire."[68] The famous passage about Abraham and his son Isaac dramatically peaks with the nick-of-time appearance of a messenger from God, who brings a ram that Abraham can use as a burnt sacrifice in place of son Isaac, as had first been commanded by God. This passage is usually interpreted as signaling the historical switch from human to animal sacrifice.

Direct archeological evidence for these ancient events is scarce and controversial. But recently, archeologists have obtained extensive proof of human sacrifices performed by the Moche culture of ancient Peru, just one of many ancient cultures that practiced human sacrifice.[69] And there were eyewitness accounts from Spanish conquistadors during the conquest of Mexico, of Aztec sacrifices on the pyramidal temples, with heads rolling down and still-beating hearts ripped by knives from breasts and held up to the sun.

One of the most intriguing examples, which occurred in the mythic sacrifices of ancient Greece, involved the three daughters of King Erechtheus. The youngest was sacrificed by her father's hand, her death ordered for the common good by the Delphic oracle, to save the city of Athens from its first major crisis, an attack from the enemy army led by Eumolpos. The three daughters had vowed to die together, so in the myth all three are sacrificed. As promised by the oracle, the ensuing battle went to the Athenians. There were other sacrifices of virgins in Greek mythology, such as Iphigenia, daughter of King Agamemnon, to bring winds to sail his army to Troy and begin the Trojan War, which ultimately was victorious for Agamemnon. But what makes the sacrifice of the daughter of Erechtheus so significant for us is that it appears as the central organizing theme for perhaps the greatest monument in Western civilization, namely the Athenian Parthenon.

Art historian and archeologist Joan Connelly has shown that the myth appears as the central subject of the frieze sculptures, the sculptural bands around the outside of the inner structure of the Parthenon.[70] The sculpture shows three maidens. The youngest is partially disrobed and is being handed a cloth by a bearded man. An adult woman at the man's shoulder watches two other young women approaching, carrying stools or baskets on their heads. Connelly interprets this scene as King Erechtheus handing his youngest daughter the cloth she will change into before her death. The bundles carried by the other daughters hold their own sacrificial robes, because these sisters have vowed to die as well. Death is surrounded by culture, in this case not only the grand meaning of the act and hopes for victory of the city, but even down to the humble form of a special clothing.

Neighboring parts of the frieze show sheep and bulls headed to sacrifice, usually thought to be parts of the annual Athenian festival called the Panathenaia, in honor of Athena's birthday. But Connelly thinks that on the Parthenon this birthday idea makes less sense than the interpretation that the sculptures show the very first Panathenaia, commemorating the original sacrifice of the young daughter as founding epic of the city, and a metaphor for the victory of Athens over the Persians, which prompted the construction of the Parthenon itself. Therefore animal sacrifice was held in commemoration of a mythic, founding, human sacrifice. Even the name Parthenon has been looked at anew in this integrated assessment of what the sculptures meant to the ancient Athenians. Connelly shows that "Parthenon" is best understood as plural, and thus the temple site was recognized back then not as the place of the "maiden" Athena but of the sacrificed "maidens."

The sacrifice of girls seems misogynist in the extreme. But in Connelly's interpretation, the sacrifice of virgins was a way that girls could be heroes just like the young men who gave their lives for the fatherland in battle. In a play by Euripedes about the myth of King Erechtheus and the sacrifices, his wife Praxithea hears the proclamation from the oracle that their daughter must die to save the city. Praxithea then makes the following speech:

> *We have children on account of this, so that we may save the altars*
> *of the gods and the fatherland; the city has one name but many*
> *dwell in it. Is it right for me to destroy all these when it is possible*
> *for me to give one child to die on behalf of all?* [71]

Connelly further explains:

> *The role of women in sustaining and preserving the life and culture*
> *of the community is paramount in . . . stories presenting themes as*
> *grisly and, at first, as misogynous as that of virgin sacrifice. Yet,*
> *valiant and heroic women go willingly to death, and are proud of*
> *their vital roles in saving the populace. They are the equals of men*
> *who die in battle for their cities.* [72]

We see here how death supports life of the higher, more encompassing level, that of culture. Erechtheus and his daughters are a myth. There are no burials under the Parthenon. Indeed all Greek sacrifices of daughters are myths. There is no concrete evidence for them. But this makes such "events" all the more significant. We see how the human mind has created out of controlled death a sense of order and progress for the whole. To fulfill the idea of "death, thus life," we have shifted levels, to the ways that such mythic deaths go from the individual out into the larger social system. Here is Connelly once more:

> The fact that the royal family participates fully in the self-sacrifice required to save the city . . . must be seen as antithetical to barbarian ideology and in keeping with the "democratic" social values of the Athenians, which allowed no one family, not even the royals, to put itself above the common good. [73]

Death by sacrifice engages multiple levels of reality. In the myths of human sacrifice, the individual is meshed with the society, because the individual's death is a functional part of the progress of the whole. Indeed, one common theme of myths universally is sacrifice that flowers into consequences vital for humanity and the world.

In the Native American myth of the corn mother, she ordered her own death by her husband. [74] The land had become barren of game because too many people lived. After her sacrifice, two of her sons dragged her body across the earth, shredding her flesh. Seven moons later, corn fields were born, from which the people then ate in abundance, saving seeds for future plantings. Where the corn mother's bones had been buried, a new herb grew forth. This the people used as ceremonial smoke to elevate their spirits.

Where do we stand today with such concepts? Shortly after the inauguration of the Ara Pacis, across the Mediterranean, from the culture that honored Moses and burnt offerings, a movement arose around the life and death of a person named Jesus. This movement eventually came to regard Jesus as having been sent by God as his Son, to be sacrificed for the sake of humanity. Among his honorifics would be the phrase "Lamb of God." The Jesus movement is still widespread, and focuses around the theme of sacrifice, not by us for God, but by God for us.

In Buddhism, there is the figure of the bodhisattva. Bodhisattvas are individuals who give up the chance at final enlightenment now to help others less advanced along their own paths. The concept here is not death per se, but a giving up of individual opportunity for the purpose of helping others, meaning the larger society.

Animal sacrifice is not as it once was. For instance, early Christianity's conflict with pagan religions was generally successful at squashing rituals of sacrifice, in essence replacing them with the ideology of that of Jesus. There has been a general historical trend to desacralize the killing of animals for meat. But as we saw, a group might still kill an eagle as part of its cultural rituals. An annual ram sacrifice is still important for areas of Islam.[75] And I have often seen it stated in technical papers, for instance regarding brain research, that a laboratory animal was "sacrificed." Its brain went to science; knowledge for everyone came from the deaths of a few. Whatever we individually feel about the issue of animal rights, we see the sacrifice concept still at work in rationalizing such deaths: death of some for the benefit of the whole.

Can we look at any human death as a sacrifice because of its effects on organizing social gatherings and on the psyches of the survivors? When the Mimbres buried their dead in floors with broken bowls over the heads, each death likely served as an awakening to the others. Human deaths are not controlled by humans (in general, excepting war and executions). But biology kills us, biology has control over the act of death. Therefore, symbolically each of our deaths will be a sacrifice in that those of the larger society will be moved, and advanced in their own quest to understand life. Nobody just dies without awakening others.

MANAGING
TERROR

Funerals and ancient sacrificial rituals are ways through which death becomes social, whereby death helps form structures in the larger life of the collective society. Such rituals that continue in today's world are the result of successful trials in how to deal with the wrenching fact of death of those who are loved by others. But what about the knowledge of one's own death? Is this knowledge just a personal fact to be dealt with inside one's brain, or does it have social implications? And does society and its structures impact on how one deals with the awareness of personal mortality?

Looking within oneself for answers to these questions is an important step to take. But this "first-person" science is very limited, given how difficult it is to perceive the immense powers of the cognitive unconscious through mere introspection. Furthermore, a single individual's findings about the dynamics of death within himself or herself might be true only in an idiosyncratic sense. To make generalizations about the relationship between the awareness of personal mortality and the surrounding body of culture, we need experiments run on statistically relevant numbers of people.

Some social psychologists are, in fact, boldly going beyond the limitations of self-examination and introspection. These scientists are conducting world-class research into the psychology of death. They find that our specifically human responses to the knowledge of death permeate our psyche in profound ways that would scarcely be guessed at during moments of ordinary thought.

I first became aware of this research at the 1998 meeting called Toward a Science of Consciousness. Held every two years in Tucson, Arizona, it features some of the leading figures in consciousness research, who present to large plenary audiences their findings about such subjects as dreams, vision processing, and human cognitive evolution. The meeting also convenes concurrent sessions for specialty topics, with perhaps a half dozen running every afternoon. That year I saw in the abstracts one session that promised talks on something called "terror management theory." The terror referred to the terror derived from our ever-present knowledge of our own mortality. The theory was entirely new to me, so I decided to check out the session.

The session was run by scientists enthusiastic, dedicated, and bright. They had been working together for years, conceiving experiments as a group and co-authoring papers with explicit statements that they shared equal responsibility for the research, despite the ordering of authors on the paper. Moreover, it soon became apparent to me that they were onto something very big. Leading off the session, Sheldon Solomon, a psychologist from Skidmore College in New York State, presented their group viewpoint that all discussions of consciousness should present questions that deal less with how we tell something is red when it is red and more with how aspects of consciousness serve deep emotional needs. He also judiciously criticized much of the conference for disregarding social parts of consciousness. By contemplating death, he said, terror management theory directly addresses the connection between personal consciousness and society.

The other two central researchers of terror management theory are Tom Pyszczynski of the University of Colorado in Colorado Springs and Jeff Greenberg of the University of Arizona in Tucson, also professors of psychology. This triumvirate has recently been joined by Jamie Arndt of the University of Missouri and Linda Simon of the University of Arizona, plus a growing number of others, including colleagues around the world performing experiments in other nations. I will describe some of their extraordinary research results in a moment. But first it is necessary to lay out the background of their ideas.[76]

They observe, first of all, that animals are designed to survive and reproduce. Behavior is directed toward these goals, via precise sensory and muscle responses, as well as judgments and habits formed by hundreds of millions of years of animal evolution.

We humans, so successful at spreading everywhere and manipulating nature for our survival, have succeeded in satisfying this urge to live and reproduce to a superior degree. Many of our special skills at survival and propagation derive from our large brain and all the mental acrobatics it facilitates. Somewhere in human evolution, a simple, symbolic, thinking process, which was so good at helping survival and was based to a large measure on the ability to make projections into the future, started to realize the truth— not only do other humans die, but "I," being human, will also die.

The realization of one's inevitable death, born from the magnificent brain, conflicts with the functional purpose the brain has to survive. This realization reveals that someday survival will not be possible, no matter what acrobatics are performed, no matter what foods are eaten, no matter what pyramids are built, no matter what chants are intoned, no matter what intensity of mental effort is applied to try and figure out some way to continue going along with life as it is. Today we confront this fact perhaps more poignantly than ever, because medical advances often extend life immensely—but not indefinitely.

Thus there occurs within us a tremendous collision, between our will to survive and our knowledge that it will ultimately be thwarted. I like to call this collision the "primal clash."

The psychologists of terror management take the primal clash as a psychological given. Then they ask what we do about it, because it can cause "abject terror." But usually we feel all right. So most of us most of the time are buffering this deep anxiety. In terror management theory, the simplest ways that the fear is controlled and overridden constitute what are called "immortality projects."[77]

These promise immortality in the face of the blatant and disturbing knowledge of mortality. Some immortality projects are literal. In them actual immortality is promised. A premise in the primal clash is thus negated. No—all humans do not die. In this category we find the many types of afterlife scenarios, projected by different cultures throughout history, some of which are still very active today. The promise of immortality occurs, for example, in the final line of a bedtime prayer I knew in childhood:

Now I lay me down to sleep,
I pray the Lord my soul to keep;
If I should die before I wake,
I pray the Lord my soul to take.

According to terror management theory, there are other forms of immortality projects that are purely symbolic. These operate through identifying and seeking ongoing life in one's words, or the spread of one's personality and beliefs. Seeing your progeny carry on, or writing books, or thinking, for example, that one's words will resound in future years, are forms of symbolic immortality projects.

The psychologists of terror management are going much further than the concept of immortality projects in probing the structure and function of death-denying defenses. In the concepts of a literal afterlife and living on via symbolic immortality, it is fairly easy to see how these would help counter the terror of the primal clash. But as the psychologists are discovering, in buffering our anxiety from the primal clash, we utilize a number of psychological mechanisms that at first blush have nothing logically to do with death. We employ these mechanisms in daily life, in ways that deeply affect the connection between society and the individual self. By these mechanisms our personal death as a concept in our minds becomes transmuted into defensive structures of our lives. The most studied of the two primary mechanisms so far involves what are called cultural worldviews.[78] Cultural worldviews are "humanly created and transmitted beliefs about the nature of reality shared by groups of individuals."

What is a cultural worldview? It's the view that Catholics have the correct take on the nature of God. That Buddhist meditation is a good way to enlightenment. That hair dyed green is the ultimate in cool. That the Yankees will remain unbeatable next year. That particle physics will give us the final answers to the universe. That capitalism is the best way to upgrade living standards all across the world. That the albatross is the most crucial sea bird to conserve with all-out efforts right now. That rock and roll will never die. Cultural worldviews are the brickwork in our edifices of politics, religion, economics, law, art theory, science debates, sports, fashion, diet fads.

There is somewhat of a fuzzy border between a fact and a cultural worldview. That DNA has four kinds of bases is obviously a fact and not a worldview. But DNA can become an object within a worldview when people believe in something more than an objective fact. For example, one could support the opinion that DNA analysis is the royal road to a terrific human future. A worldview is when a feeling is held about some characteristic of the world for which someone else could well have a quite different feeling.

Note that cultural worldviews are, by definition, not strictly private. They are worldviews because they are shared by many people. Furthermore, we as individuals who participate in the worldviews know that our view is in fact shared. Crucially, however, the view is not usually shared by all. This creates what sociologists call an in-group of those with the shared view. Others constitute the out-group.

Another characteristic of cultural worldviews is that they are, to some degree, arbitrary. They are human creations. It is sometimes difficult to swallow this fact, but a sober look across human geography and history shows it to be true. It's not that the views lack practical functions in the times and places they exist. Nonetheless, they do change over time and often differ within social units across scales that range from family and neighborhood to nation and the world.

Holding cultural worldviews seems like the ordinary bread of everyday, everyman-and-everywoman human existence. However, terror management theory takes worldviews as not simply obvious, but as human mental attributes whose root functions need to be profoundly questioned. Why do we hold worldviews? The answer of the terror management psychologists is that this human attribute at least partially, and perhaps largely, functions to help us live with the primal clash. The mechanism operates so smoothly as a system of psychological defense against the terror of the knowledge of death that we do not even suspect it functions as such.

How do cultural worldviews specifically function? According to the social psychologists of the theory, the worldviews "assuage the anxiety engendered by the uniquely human awareness of vulnerability and death. Cultural worldviews ameliorate anxiety by imbuing the universe with order and meaning, by providing standards of value that are derived from that meaningful conception of reality, and by promising protection and death transcendence to those who meet those standards of value."[79]

These are all fine-sounding concepts. But where is the pudding of proof? The demonstrations of terror management over the years have sprung from a hypothesis derived from the above reasoning. It is called the "mortality salience hypothesis":

> *If a psychological structure provides protection against the potential terror engendered by knowledge of mortality, then bringing thoughts of mortality into consciousness should increase concern for maintaining that structure.*[80]

This hypothesis predicts that if the worldview or system of worldviews that people hold does in fact serve as a buffer against the terror of mortality, and if reminding them about their mortality will be experienced as an inner threat that could bring forth that terror, then, as result, people will strengthen their worldview. In the experiments, this strengthening is sought by measuring the extent to which subjects can be induced to more adamantly either support those who share their worldview or denigrate those who do not share it. This excess of support or denigration is called "worldview defense." If the worldview functions as predicted by the mortality salience hypothesis, then worldview defense should result from experimental manipulation of the awareness of mortality.

In the experiments, how is mortality made salient? Very delicately, with just the slightest prick to consciousness. The most common technique used in the experiments by Greenberg, Pyszczynski, and Solomon uses a couple of questions embedded among others in a personality questionnaire. Subjects fill out this questionnaire at the start of a typical experiment. They are told nothing of the true purpose of the overall experiment. Instead, they might be informed that it is about personality traits and interpersonal judgments. Hence the questionnaire to start.

The crucial, embedded questions request half the test subjects to "Please briefly describe the emotions that the thought of your own death arouses in you," and "Jot down, as specifically as you can, what you think will happen to you when you die and once you are physically dead."[81] The other half of the subjects serve as the control group. These control subjects do not undergo the mortality salience of the special questions. Both groups share all the other personality questions on the form.

The next phase of the experiment includes the actual test that will be measured. In many cases, the psychologists use sham essays about America. These are written by the psychologists but are purportedly by foreign students studying in the U.S., published in a bogus political science journal.

Subjects in the experiments have to read these essays and then rate how much they like or dislike the authors. One essay is weighted very heavily to be pro-U.S., basically with the message that the U.S. is the greatest country in the world and the land of opportunity and freedom for everyone. The other essay is bitterly anti-U.S. Its dispatch is that although the U.S. postures as the fighter for democracy and the pursuit of happiness, the reality is a country in which the rich become ever wealthier and the poor more down-trodden, and that the country's ideals are no more than an advertising facade.

The findings of a typical experiment are as follows. On average, subjects in both groups rank the pro-U.S. essayist as more likable than the anti-U.S. writer. But the ranking is distorted by the subjects whose potential for inner terror had presumably been perturbed a bit with the questions about their own deaths. The group who underwent mortality salience rank the pro-U.S. essayist as exceedingly likable and the anti-U.S. essayist as exceedingly disagreeable, compared to the ranking from the control group. This accords with the prediction of the mortality salience hypothesis.

According to the theory, the subjects who are made salient of their mortality will need to bolster their inner defenses against their inner terror, and they do so by defending their worldview. This is what is found. These subjects praise those who are similar to themselves (assuming, of course that the students are themselves pro-U.S., generally a good assumption in the last couple of decades without, say, any Vietnam War). And they denigrate those who hold opinions different from those of the mainstream values of their own worldview as U.S. citizens. It is important to note that the students do not necessarily feel abject terror consciously, then relieve it by defending their worldview. No, the idea of terror management is more subtle. The students are already immersed within the dynamics of their unconscious, which creates the fairly pleasant worldview in ongoing daily life, continuously suppressing the disturbing idea of death. If death awareness is awakened just a bit by the death questions in the questionnaire, the students are not even aware of its awakening. Nor are they aware of how the worldview defense mechanism kicks in more forcefully than normal. This all happens in the unconscious, as will be described further.

An important criticism was leveled by other professionals toward this and similar findings in the early days of the theory. Namely, how do we know it was the idea of death and not the general anxiety caused by the mortality salience that caused the responsive defense of worldview? In other words, was the distorted evaluation of the essayists by the test group of students due to their state of anxiety?

If the criticism is valid, then other ways of provoking anxiety should give the same results. To examine this issue, Greenberg, Pyszczynski, and Solomon inserted into the overall experiment an additional questionnaire, after the introductory personality questionnaire but before students read and evaluated the pro- or anti-U.S. essays. This new questionnaire measured anxiety, and had previously been shown by others in the technical literature of social psychology to be a solid metric of the intensity of anxiety.

Significantly, the group of subjects that underwent mortality salience did not exhibit elevated levels of anxiety. That means that merely answering a couple of questions about death during a routine personality questionnaire did not create anxiety. Additionally, experiments were run to provoke anxiety and then to see what happens to worldview defense. In those slots on the personality questionnaire where the questions about mortality occurred, the psychologists inserted new ones aimed specifically to create anxiety. For example, the subjects might be asked to describe their feelings toward final exams, or about public speaking, or about their eventual job prospects after graduation—all stress-inducing issues for students. When the results were tallied, questions about exams and jobs did indeed elevate the level of anxiety in the subjects. But—and this is the crucial point—these subjects, whose anxiety had been experimentally elevated but who had not been asked to write about their death, did not exhibit worldview defense in rating the pro- and anti-U.S. essayists. Thus anxiety itself does not trigger the reaction of worldview defense. But being reminded of one's mortality does.

The implications are that worldview defense is a specific antidote against anxiety engendered by awareness of one's mortality, and that the defense is so well-constructed that under normal circumstances anxiety is controlled to a level that we do not even notice. Before anxiety arises from anything that triggers the awareness of mortality, worldview defense kicks into action. And it's all unconscious.

These essential findings have been replicated and improved by countering other alternative explanations over the years. Overall, in more than eighty experiments in both the lab and out in the field, and in other countries such as Canada, Germany, and Israel (with more in progress elsewhere), the theory has stood up, strengthening via replication and developing complexity as findings from different experiments start to dovetail into a system.[82] For example, some of the field tests involved questioning pedestrians about a hot political issue—immigration laws, in one German experiment—which then showed that responders exhibited more worldview defense when the interview took place in front of a funeral parlor, rather than a hundred yards away.

The psychologists tested actual Tucson judges who volunteered for an experiment. The judges were given legal briefs of arrests for charges of prostitution, and were then asked what bail they would normally set in such cases. Half had been pricked with mortality salience (using the death questions in the "personality" questionnaire) and half were controls who had answered the same questionnaire except for the death questions. Bail set by the judges after mortality salience was nine times higher.

The same experiment was run with students acting as judges, yielding the same results. In the case of the students, they could also be questioned beforehand about their feelings toward prostitution. Only those students with negative feelings about prostitution set statistically higher bail for the cases. This showed that the bail was not just a matter of being in a bad mood and lashing out at any convenient victim after pondering one's mortality. Instead, for the anxiety check to work, one had to actually defend a real worldview. If one accepted prostitution, there was no anxiety block to be gained from setting a higher bail for an alleged prostitute.

In yet other experiments, students who had undergone mortality salience bestowed higher rewards for heroes. This showed that the defense mechanism was not necessarily tied to punishing those in one's out-group. It could operate as well by supporting those in one's in-group.

In some particularly fascinating work, the psychologists have demonstrated the importance of the unconscious in these dynamics. For example, if the test subjects are instructed to think very carefully about why they are passing judgment, say on the pro-U.S. and anti-U.S. essayists, then the distorted ranking toward in-group and against out-group does not occur. Making the process conscious does not allow the unconscious to perform the repression of death anxiety by the seemingly irrational mechanism of worldview defense.

The power of the unconscious is further shown by the fact that the induction of mortality salience itself does not have to be conscious, even for a brief period of time, which it is, say, in the case of writing an answer to a question about one's body after death. (The subjects have mortality in consciousness, briefly, while writing the answer, but of course they do not know why they are being asked to answer such a question.) To eliminate consciousness altogether, the psychologists used subliminal words. Mortality-reminding words, such as "death" or "dead" were quickly flashed on a screen, sandwiched between other neutral words that were on the screen long enough to be readable. To the subjects the sandwiched trigger words of death reminders were invisible. But despite this invisibility, worldview defense was still brought on by the subliminal "death" words. Thus the words must have been seen somewhere in the brain by the unconscious, which subsequently cranked up its worldview defense mechanism, as measured in the tests, using, for example, the pro- versus anti-U.S. essays. Neutral subliminal words of the same length, such as "field," did not trigger the defense. Ah, but what if another potentially disturbing word such as "pain" were to be flashed as the sandwiched trigger? This was tried, the results were negative, as predicted by the theory. Only "death" seems to spark the worldview defense!

A second psychological mechanism, though less studied, has also gained an impressive amount of experimental support. It involves our urge to enhance our self-esteem.

Some of the experiments on how enhanced self-esteem can buffer the anxiety from pondering mortality have been modifications on the tried-and-true experiments on cultural worldview defense. The basic idea is that if either self-esteem enhancement or worldview defense can repress the anxiety that might arise from mortality salience, then subjects who were given boosts to their self-esteem should not need to defend their worldview. This is indeed what has been found.

For example, the self-esteem of subjects can be manipulated early in the experiment by having them undertake a word puzzle. At the end of the allotted time they are graded. No matter what their actual accomplishment, they can be told that they scored either in the upper echelon or bottom rung. Those who were told they were top dogs sally forth into the next phase of the experiment with enhanced self-esteem. The others have their tails hanging in shame. Now the experiment continues on with the mortality-inducing questionnaire and then finally, for instance, the reading and ranking of the pro-U.S. and anti-U.S essays. It is only the subjects with lowered self-esteem who distort the ranking and thus use worldview defense. Subjects with the artificially induced, high-flying self-esteem merely ranked the essays the same as the control group did. Thus a high level of self-esteem by itself functions as an anxiety buffer with no need of additional buffering using worldview defense. The high flyers who underwent mortality salience had been given an opportunity for worldview defense. They did not need it.

Self-esteem and cultural worldviews are clearly related. With cultural worldviews we see ourselves as part of a larger shared system of specific beliefs and values. With self-esteem we see ourselves as a valued member of the in-group with a specific worldview. It appears from the experiments that feeling better about one's status within the in-group (say of intelligent people) can serve the same anxiety-suppressing function as defending the cultural worldview itself, say by denigrating members of the out-group.

The experiments with self-esteem can focus on the human body, because for all of us the body is a source of mortality salience. If there's one thing we know about death, it's that our bodies age and undergo eventual, terminal demise. Being reminded of one's body can be a source of lowered self-esteem, since awareness of our raw, animal presence is a message about mortality.

This twist in the theory was tested by Jamie Arndt.[83] All the test subjects wrote responses to the personality questionnaire with the standard two death questions. (Remember, in previous experiments only half are asked the death-related questions.) The difference in this experiment between test group and control group was in the decor of small, private rooms where each subject was seated to fill out the questionnaires. Half the students composed their answers in rooms that had large mirrors on the wall in front of their writing desk. These subjects could not avoid at least occasionally glimpsing themselves in the mirror. The other subjects in the control group wrote in rooms that were identical in all ways except without mirrors.

The theory of terror management postulates that because self-awareness is necessary for the idea that we exist, it is also necessary for the primal clash, which requires the concept that someday we will not exist. And being faced (literally) with one's physicality is a worrisome threat to dignity. Therefore heightened self-awareness and feedback on physical presence should increase the possibility for calling up in the unconscious the inevitability of one's death and thus feeling the terror of the primal clash. What Arndt measured was how long the subjects took to fill out the questionnaires with the two death-related questions as well as about twenty more, all requiring short answers. Did the mirror make a difference?

When they were finished, subjects came out of their rooms to start the next phase of what they thought was the real part of the experiment. Other, bogus tests followed. But as far as the psychologists were concerned, the experiment was over the moment the subjects stepped out. According to the theory, subjects in rooms with mirrors, in facing answers to questions about death, would be less able to face themselves in the mirrors. So they would more quickly write and thus more quickly exit the room.

Here are the specific results. Subjects who sat in the rooms with no mirrors took a little over six and a half minutes to complete the questionnaire. But subjects who sat in rooms with the mirror took nearly two minutes less.

The same issue that was raised in earlier experiments was asked: Were the results a combination of the mirror plus the anxiety that was possibly created by the death questions? No. This was shown by another experiment in which some students had no death questions on the questionnaire but were required to answer questions about taking a final exam. As determined earlier, this line of questioning had been proven to produce measurable anxiety. But answering these anxiety-producing questions in a room with the mirror or without the mirror did not affect the total times spent on that questionnaire. Thus the time difference originally found was not caused by the mirror plus anxiety per se, but by the mirror plus the specific contemplation of death.

What are the implications of all these remarkable findings of terror management theory? We can now revisit our axiom of "death, thus life" in a new sense, a sense not even dreamed of prior to these experiments. Death not only affects rituals that bind culture. Death also influences and thus gives life to our daily, pervasive mental structures by which we are bound together with shared worldviews and methods of supporting each other's self-esteem. Complex aspects of our psyche, aspects that might be thought independent from the concept of death, are at least partly and perhaps substantially forged by the presence of the worm of knowledge about mortality. Our very psychological life and the way we share reality with others is partially sculpted as a response to death. As biological death is turned round into new life forms via carbon flows in the biosphere, so human death is turned into social and psychological structures via loops between individuals and their cultures.

The theory can provide us with new insights into the origin and maintenance of human aggression. An ultimate way to defend one's worldview is simply to eliminate the others, thereby proving that "my" worldview must have been the right one.[84] As Sheldon Solomon noted, he could not run experiments with subjects given opportunities to use flame throwers and hand grenades as a way to test worldview defense. But he did once use a killer hot sauce as "punishment." Subjects first went through a standard battery of terror management questionnaires, then moved on to what they were told was a second complete experiment. There they were to allot an amount of the hot sauce for a taste test to a participant from the "first" experiment with either similar or dissimilar political views. Those in the group with dissimilar views were "punished" with double the amount of hot sauce, compared to the amounts allotted those of similar political persuasion. This was only true for the allotters who had gone through mortality salience. Therefore, the theory says that wars and other conflicts (perhaps in day-to-day business) partly serve to soothe the disturbing rattles of peoples' primal clash, which can come to life whenever differences in worldview become known. These differences threaten the respective worldviews, a process which thereby decreases the level of protection provided by them and allows the abject terror of knowledge of mortality to rise up. If you can't persuade them to join you, then annihilate them.

But how often are normal people made aware of their mortality? We are not every day writing answers to questions about our death. Nevertheless, the tests in front of funeral homes show that just subtle reminders of death are all that are necessary.

Searching for some mortality salience? Pick up a newspaper. Turn on a television movie.[85] If death motivates our unconscious according to the mechanisms found by terror management theory, we are primed to defend a worldview and create bias toward others all the time. Need more? Look in a mirror, have sex, or defecate. Studies by the investigators of terror management continue to show that reminders of our animal (and thus mortal) natures are enough to trigger the enhancement of worldview and struggle for high self-esteem.[86]

Because everyone holds a variety of cultural worldviews and is faced with the fact that others don't hold those views (or are often supporting the opposite views), we experience threats to our defense systems all the time. Thus we are tempted to denigrate. One aid for this situation has come from the experiments. Tolerance itself can be a worldview. When the psychologists took subjects with extremely liberal values through the battery of tests of terror management, the subjects did not need to denigrate their worldview rivals—conservatives for example.[87] In fact, after mortality salience, liberals liked conservatives more, presumably because they were in essence defending their own value of liberality by putting it into action in being even more tolerant. Thus the theory of terror management gives educators and social scientists the means to study tolerance in a new way. We can do harm by our opinions that are not true judgments but prejudices, powered by the armor they provide against knowledge of our own death. If we can be aware of this process, then death as a motivator in our unconscious could be put to use, theoretically, to create a more just and tolerant society. The energy in wood can be used to burn down a town or to power a steam engine and make electricity. In other words, we might not be helpless in the face of the psyche-sculpting power of death awareness, but rather seek to use the power to create socially desirable reactions in individuals.[88]

To repeat: the tolerant can be more tolerant when faced with the awareness of death. The bigoted become more bigoted. It's a choice we need to think about. I have found that the more I cultivate gratitude the more gratitude is created. What worldview is drawn from to defend against the anxiety of death awareness matters for the future of global society. We should study modes to develop self-esteem in ways that include everyone, and we might eventually be able to harness the power of death awareness to this regard. Indeed, we might have little choice but to use death awareness as an important component in our quest to make a more desirable society.

In addition, the theory may provide archeologists and theoreticians of the evolution of culture and consciousness with dynamic new questions about how the awareness of death related to culture in the human past. For example, the theory could look at a big question in cultural evolution: Which came first, culture or the awareness of death?

One approach might consider a leapfrogging of the two, in stages whose lengths may eventually be traceable in the archeological record. Culture, say in the material form of stone tools of *Homo erectus* a million years ago, could have existed before the awareness of death, because stone tools (as culture) directly supported the survival and reproduction of those early hominids. Only later the awareness of death would have dawned, within the larger brain evolved, in part, by the successful and increasing use of cultural survival tools—at first material but then increasingly symbolic. Then this adaptive mental phenomenon of culture, as manifested in more advanced behaviors, could have provided a new function, namely, to help deny death through shared cultural worldviews. By buffering the idea of personal death, culture would then have had a brand new function and thus reason to be driven toward still more complexity, including the ritual of early burial that removed the dead from campsites. Perhaps new cultural practices first honed through the symbols and rituals around death awareness were then transferred to other domains of life. And so on, as culture and death awareness as a cultural power reinforced each other in a spiral manner. This reinforcement could have propelled overall cultural evolution.

We can all be considered as wounded by the clash between our urge to live and our awareness of eventual death. Culture is part of our salve and our bandage, a bandage we probably cannot remove. Even if culture were solely part of a defense mechanism against facing our awareness of the inevitable, we could not fling it off because it is too late—we cannot return to non-awareness. Let us therefore be thankful for what we have. We can be grateful for who we are as conscious beings in a primal clash that is partly resolved by the cultural web around us. As tendered by the terror management group: "One provocative conclusion that can be drawn from the evidence available to date is that by motivating adherence to culturally prescribed beliefs and values, the dialectical opposition of an animal instinct for self-preservation with awareness of the ultimate futility of this pursuit may be what makes us, for better and worse, most distinctly human."[89]

Overall, however, just to throw some cold water on this final, appreciative swaddle of warmth, there are disturbing aspects to the robust experimental findings. The findings do prod us to personally probe how our self-esteem and worldviews are at least partly what they are because of their unconscious functioning in shielding us from the awareness of death. Aspects of life we do not question because they seem so natural can now be deeply questioned as not being all they might first seem to be. For instance, is singing the "Star-Spangled Banner" partly a cover-up of death awareness? Maybe we should become more aware of death and not need so much shielding, and so many battlements against living with such awareness. But then do we chance becoming more disturbed? Terror management theory provokes us to question the usually unquestioned. Why do we seek what we seek? Why do we do what we do? Why do we believe what we believe? As we reconnoiter these issues the path will not always be the bliss of ignorance but sometimes the pain of knowledge.

I, myself, have been changed by one episode in particular during this study of terror management principles.

For several days after an especially intense round of reading papers about the research results, I found myself reviewing the memory of an event during my childhood. I was about thirteen years old at the time. I was reading one of the books in the Time-Life Science Library. I loved those particular books because they held fascinating text and sumptuous pictures. They were rich fertilizer for my brain. This remembered book was in fact about the brain. A section about memory dealt with intelligence and a metric of intelligence called the intelligence quotient, IQ.

The illustrations showed pictures of famous people, all now long dead. Along with their images, IQs were boldly put forth in ink. Their IQs had not been measured with an elaborate version of the twentieth century's Stanford-Binet test, which I had taken a couple years earlier for possible entry into a college preparatory school. Rather, psychologists had estimated (obviously correctly, in the mind of a thirteen-year-old) the IQs of these famous personages. Included were people familiar to me, such as George Washington. Others I had never heard of. I already knew my IQ from the prep school test, and was overjoyed to see that mine was higher (i.e., that I was smarter) than Washington himself. Wow, I thought. That meant I could be as famous as he is, that I would probably go down in history, that I would someday have my name and picture in a book such as the one I held.

True, my rank paled before that of someone named Voltaire. But who was he, anyway? And what about this guy Goethe, who also boasted a number well above mine. Well, whoever they were, they must have done something great to be in the book. Considering my friends, who all had heard of George Washington but not Voltaire, I concluded that my being closer in IQ to Washington than to Voltaire or Goethe was more than good enough for me.

What a most excellent thing it is to be renowned in a book, I thought. And today, in reviewing these memories, how surprised I am at how clearly I still see myself looking at that gallery of IQ notables. What security I gained and what self-esteem I enhanced by seeing that my brain's capabilities were situated within the range of those on that fateful page. What immortality I then saw ahead for myself. All that was needed was for me to do something famous. But that was, as I felt then, basically in the bag, or in the brain.

It has become surprisingly clear to me that I have been using this incident, ever since I was a young teen, and others like it, as part of my personal terror management. By that age I was fully aware that death lay ahead;

this awareness, along with my self-esteem in a world with a huge variety of competing worldviews, was likely an issue that concerned me.

Contemplating all this has made me realize that one of my goals in life has been fame. It has not always been conscious, and I have not done a very good job at achieving it. But as an unconscious goal it has been a deeply-seated driver in my brain. I have not been interested in money (or hardly, anyway) or in personal possessions. But fame, yes. And it traces back to a personal sounding at least as fathomable as my memory with that book about IQs.

Why would I be interested in fame? Could it lead to having money? To having girlfriends? Possibly it would have these practical functions. But I now feel clearly that I was using it as a way of grasping at immortality in the face of obvious mortality. Those high-IQ guys took hold of my mind as I gazed at their mugs in the book. And, to compound the problem caused by my desire for fame, I have a worldview that conflicts with the urge for fame—the view that I gained later, in my twenties and reinforced many times since—a view about how to live, namely, that striving for fame is worthless.

Remembering the intensity with which I gazed longingly at those famous faces with the high IQs, I have realized, however, that despite subsequently taking up the slogan that "fame is worthless," down deep the quest for fame still motivates me. This clear, recent realization has deflated much of my quest. In the days following my contemplation of the memory and seeing how it formed the basis for current motivations, I was mildly depressed. I walked the streets haunted by this visceral desire for fame, which I had dredged up from my unconscious. I told friends, almost with embarrassment, that I had discovered that I want to be famous. (Bless one friend, Marty Hoffert. When I told him he just quipped, "What's wrong with that?".) In fact, I almost felt that after this book I would not write any more, so profoundly had the dredging started me to question my motivations for accomplishment.

My message is that the examination of death and the mechanisms by which we incorporate death awareness into our lives might lead to surprising realizations far afield from the last sting we felt at the loss of an admired celebrity, friend, or relative. We have the opportunity to turn the findings from terror management theory into a new vision for "death, thus life." Here death is the awareness of personal mortality, and it is life as we now live it that incorporates such awareness. Because many of our psychological structures that deal with this primal clash operate unconsciously, the life that death awareness creates might not automatically be the most desirable. But knowledge of the theory gives us choices as we watch ourselves search for self-esteem and take on cultural worldviews. It might be best, for example, not to cling to certain desires that are, in part, reactions to death awareness, desires such as that for fame.[90]

In conclusion, the psychologists of terror management have shown that some of who we are is built on a response to the fear of death, a clash between the need for life and the knowledge that death will thwart this need. Our response to this primal clash causes us to construct and thus inhabit a psychological world that is necessarily life on a larger scale, because it must encompass both the need for life and the knowledge of death. We therefore exist in a human nature that is some sort of resolution to the clash, a life that includes the simple desire for its continuance and the knowledge of eventual death.

The visions revealed by terror management theory can themselves become a kind of culture we can learn to live within, a kind of meta-culture, or a worldview about worldviews. And this going beyond can lead to liberation through knowledge, knowledge that the awareness of death drives society and its individuals in ways scarcely perceived. Why liberation? The path is yet to be determined, but we can be sure that with new knowledge comes new potential for life. Realizing that death structures our life in the present can lead to a new kind of future life, a life more conscious of the interpenetrating co-existence of death and life.

DEATH WITH INTERCONNECTED DIGNITY

Deaths of individuals are very much tied into the social body. This is true in ancient sacrificial rituals and in today's funerals. It is even truer in the way that our anxiety about mortality is transmuted in our psyche into support for worldviews, as uncovered by the experiments in terror management. In general, how should we consider the relationship between the individual and society? The social body is literally a body of individuals engaged in a psychological symbiosis. When we as individuals start to see ourselves as part of a larger body and, indeed, see that aspects of the larger body are within us, then we have what I will call the experience of the extended self.

One of my own first conscious experiences of the extended self took place in June of 1984, at the funeral for beloved grandmother Babci. Standing before the open casket during the evening of "viewing," in the midst of mourning by family and friends, I realized that some of my siblings had a different view of what her death meant. They referred to Babci's presence among us at the funeral home, hovering as a spirit, an individually conscious and disembodied soul who was watching us watch her former body. In tears, I had to admit to myself that was not what I believed. And yet I could honestly nod in agreement that I felt her presence, too. It was in all of us.

For me, Babci lived in the multiple parts of the psyche of family and friends influenced by her. This distribution of her into portions of the survivors' brains did not, of course, happen at the moment of death; she had been percolating into us during the years that we knew her. But it was at that moment that I quivered with this new knowledge that we are not the individuals we think. Rather, we are composites of people we know and love.

The interconnections among people are recognized, of course, to some extent by everyone on a daily basis. But sometimes the understanding swells to a crescendo. One function of a funeral is to intensify the swelling in everyone, to shake them with the realization that threads of being interweave them with the deceased, and to burn into them how the deceased (for example an elder) lives on via what was transferred from elder to younger in the mysteries of psychological development. Babci's body was going to disperse slowly in the soil she was buried in. But her psychological being had been dispersing for decades, as leaves dropped from her into our minds, from which we grew. Now there would be no more leaves, but the presence of those nutrient packets, already in me, suddenly became magnified in her death.

Her death also propelled me into a state of mind that I now can see was an exercise in terror management. In the weeks afterward, I became obsessed with the idea that the world needed a sole replacement for its many religions. Although some are very widespread and claim to be global, every single one developed in a world that was not itself global but local: Judea, India, China, Arabia, and so forth. Today the world is finally and undeniably all-connected (so my thinking went), and so needs a cosmic outlook born from the global context. That outlook can be provided by science.

I made notes for a book about the replacement of religion by science. The book never made it past these plans. I was much too busy as a newly minted scientist. Just a month earlier I had received a Ph.D. for my research into the global cycle of carbon. But what about my obsession during those weeks, immediately following Babci's death? By postulating that my own worldview and that of my professional in-group should become everyone else's, was I having an episode of "worldview defense"? Was I fending off thoughts of my own mortality, which had painfully oozed into near consciousness at my grandmother's death, in planning for world domination of my own ideology? From the viewpoint of terror management theory, it sure feels like it.

My retired friend Ann Marek recently told me of how aware she has become about the dead being alive today, in us. As she ages, she is ever more actively contemplating the meaning of life and death, and is glorifying in the fact that Jesus is alive in us through images and teachings, that Buddha is alive in us through images and teachings. We could include any of your favorite teachers from the past. There is less separation between them and us than we think. To the extent that we share ideas, values, beliefs, goals, and knowledge, they are us. They inhabit, and thus form parts of, our consciously thinking selves and our vitally active, unconscious selves. The people in our past, including deceased friends and relatives, live on not just in us but in others who knew them, either personally or through their works. And parts of our selves are also in other selves today. It's a psychological web. This is the extended self, not just the dead in me but the dead in many of us, which thereby connects us to a body with various harmonics of extension—Jesus harmonics, Buddha harmonics, mother and father harmonics.

No doubt it is easiest to feel this extended self when someone is dead, because their language, their teachings, their lessons continue only as self-generated mental dialogues and ethical behaviors in the absence of the originator's living, biological body and self. Of course there also remains the crucial, external corpus of books and videos for seekers who follow particular teachings of those who have died. In fact, often we see that the death of a family member escalates the intensity of the key values they taught us. And death commonly amplifies the fame of the already famous. Until their death we are not aware of the full extent to which people affect us.

But the extended self is an experience created through those in the present, too. Living people affect and interconnect with each other right now, all the time. The passing of states of being between living people, however, is often more difficult to witness because of the competitive jostling that exists in the most loving relationships, in the best of friendships, as checks and balances about who the other is and to what extent everyone allows others to expand their power and general presence.

This behavior is just basic primate social skills further amplified through human evolution. It is what we are evolved to do, to slip our two cents of story and often foolish reason into conversations, into determining states of reality for others. Our unconscious makes sure that it spreads itself out into the world to be heard and admired, that our personal worldview is defended by others day by day, fitting in alongside others who are simultaneously defending their worldviews.

All this defending happens within a tremendous sea, or moving river, of humanity as a giant swell of integrated individual psyches, in which we are interconnected ripples. One particularly ideal place to see how this all works—these dynamics between the ripples as merged into the river, versus the ripples as jostling individuals—is in the practice of science.

———————

Ah, science. Sometimes the word chimes within me like the call bell of a religious order to which I swear allegiance. Some friends more attuned to the humanities said I narrowed my mind by joining that order. But instead I find myself entering into a deep communion with others, reaching out into a scientifically extended self. What an arena of human endeavor is science, a river of knowledge, ever expanding and built by collectives of individuals in widespread guilds of expertise, in which the members unite by communicating their findings.

I said "particularly ideal" because of the way that science, as a practice, has instituted the extended self into a formalism for finding and laying down truth, a praxis in which individuals are turned into cooperative parts of a larger, social organism, and by which they achieve an institutionalized immortality.

Among my first inklings of the extended self of science is the one that came to me on an autumn hill in New York State, just after a stint as a summer researcher at a NASA center in California. It was a couple years after my first obsession about replacing religion with science. Out for a hike on that sunny day, I thought back upon the summer, about the new people I had met and with whom I had been thrown into work. In this case the science team was trying to figure out how life support systems could be designed and built for human colonists on Mars or in other space colonies.

Some of the team was located at a California NASA center, with the rest situated at the home bases of other NASA sites, at perhaps a dozen universities, and in some international agencies. At one point that summer most of the key researchers gathered at the California center for several days of presentations of their work. To the collective whole they showed their contributions to the overall intellectual and engineering effort.

This summer experience, these new colleagues—some of whom I knew only through their papers or from the word of others—had influenced my thoughts about how I spent my passionate time on technical matters, who I was. I had become entwined with these folks in far-flung links around the globe, with gossamer mind fibers stretched out to thread us into a guild with a common mind. Each of us represented the organs—heart, lungs, muscles—of a social body that was trying to take steps toward understanding and capability. Before the electronic world wide web, before specialized chat groups and shared global information—which now makes the mind network of the extended self clear to everyone—science had woven neuronal electricity into its own vast world wide web.

The institutionalized aspect of the extended self of science, however, is more than the sharing of professional interest. A crucial component lies in how truth is set forth by the bigger group self. It often seems, as findings appear in the media, that some scientist makes a discovery and it is announced. In reality, the finding involves a long, sometimes tortuous process within the extended self of science. Immortality does not come without a struggle.

There are a number of tall steps to surmount. Usually a scientist first discusses the finding with colleagues in the guild. That helps him or her make sure that no obvious technical oversight occurred that would turn the finding into an embarrassing blunder. The scientist might vet the talk at a meeting, flinging forth the idea for evaluation by a live audience of the extended self. This often takes place before the scientist attempts the more serious and arduous move of submitting a paper to an archival, technical journal, which is the ultimate form that each new scientific truth takes.

Before publication the science paper must go through the gauntlet of "peer review," the key formal institution of the extended self in science. The paper is written to conform to the overall style of some specialized journal and is then submitted. A topic editor (a Ph.D. in the field) then sends the paper to two or more other experts in the field. These are likely the scientists who, somewhere in the world, are working on something very near to what this scientist has done, who can most understand every single aspect of the highly technical paper. They examine the submitted work and pass judgment upon it in a letter written to the editor. These review letters are often many pages, and the review process itself can take months of waiting.

The reviewers are anonymous. They can choose to not remain so. But the default setting of anonymity is a crucial part of the process. It means reviewers do not have to fear retribution from a disgruntled writer who intends to revenge a critical review, say one that results in blocking publication of the work. Such revenge could take place when the writer one day becomes a reviewer, by receiving the original reviewer's own written work for judgment.

What is scientific truth? There is no objective standard outside that of human beings attempting to discern the validity of another's experimental techniques, and of how the results of another's experiments expand previous knowledge and offer solutions to some puzzles about nature. Thus truth is established, step by step, within the mind network of the extended self of the scientific community. In practice, when the editor receives all the reviews, he or she then attempts to make a consensus summary based on the reviewers' opinions of the worthiness of the submission. Then the editor decides what to tell the author. The paper might be accepted, or provisionally accepted, provided changes can be made that successfully answer tough questions raised by the reviewers. Or it might be rejected outright.

When the paper is published in a technical journal, it receives permanent page and volume numbers that enable others to find it. Slices of truth can be thereby easily located according to reference code. Moreover, in writing the paper the scientist draws upon the work of others, establishing in the paper's introduction what findings have come before on some issue and what questions in the topic remain open. By presenting this introduction to the public, the new work is linked to that of the past, creating, across thousands of journals, a knowledge network of tremendous complexity. Each bit of newly minted truth does not remain so in most cases, of course, but changes as even newer work falsifies older work, enters caveats against or puts shades of variation upon it. Science is temporary, but it is not arbitrary; at any given moment the extended self is doing what it can to ensure that the latest is the best.

Science is both intensely individualistic and superbly collective. All of society is, too, but in the enterprise of science the ways that people both make their mark as individuals and work as parts of collectives are particularly clear. Thus an individual genius like Newton can then claim he saw so far because he was "standing on the shoulders of giants," the collective who came before him. The quote remains famous in intellectual history because it rings so true, placing the individual and the collective as equal partners in the enterprise. In science, self-directed people interpenetrate each other's minds and create a shared conversation of the highest form.

The writer of a scientific paper feels a kind of immortality, because of the permanence of his archived pages within the volume of a given journal. There is also a wish for everlastingness in that the writer hopes his or hers will be permanent truth. As noted, this is usually not the case, but such aspiration drives the enterprise.

I still recall the thrill when my first paper hit final ink. It concerned a tornado wind energy system, the result from experiments I conducted in a wind tunnel.[91] To be headlined in that paper with its journal number, to know it was definitive on an aspect of a proposed energy system and would have to be cited by future investigators along these lines, made me feel part of the extended mind of science, a little piece of history. I made a discovery—brought a bit of the dark unknown into the light of the known—and would be forever recorded as the man who made it.

The scientific enterprise is rather good at setting its practitioners straight about not letting their expectations of immortality run too wild.

For instance, when I take a look at the index of citations (which tracks the research that has drawn from my findings), and discover that one of my papers has not been cited for five years, it's the equivalent of a death, but without the funeral.

The microbiologist and philosopher of science Robert Pollack has written about his observations of such matters from many years of work. His conclusions about how medical science, and to a large degree science in general, functions to provide psychological solace to its practitioners appear to me strikingly similar to the concepts of terror management theory. Using that theory's language, participating in science provides a worldview to defend and a way to enhance self-esteem. Pollack agrees that science establishes a formal way to attempt immortality through discovery and publication. Science—here considered as a practice, not the findings—is an ideal worldview to gather behind as a defense, because, like many religions, it is so well defined in terms of what it is and what it does, within a tight community of fellow practitioners who believe in much the same worldview. Furthermore, as a discoverer, one fashions an individual worldview that through formal publication is disseminated to others as truth.

As Pollack emphasizes, big-time immortality in science—having your name attached to theory or, better yet, a quantity that you defined—is rare. The Watson-Crick model of DNA. Einstein's theory of relativity. Hubble's constant for the expansion rate of the universe (and the name of the space telescope!). Volta's volt. Ampere's amp. In reference to the big names in science, Pollack says, "Players in a game that can confer even this sort of immortality—however rarely—cannot be playing for only conscious stakes."[92]

Pollack means that the conscious stakes of science, such as bettering the world or meeting the challenge of solving problems, cannot be the only factors that drive scientists to work so hard. Scientists often seem to me almost sacrifices for the cause of building collective knowledge. Given that immortality of name is a possibility, the use of the profession as a way to overcome fears of death becomes, to some significant extent, a certainty. I do not claim that all those with named theories or who reached the star status of a Newton or a Darwin always had death-denying fame as motivation. In fact many had a volcanic passion for puzzle-solving. But we know that both Newton and Darwin were driven toward gaining recognition for their discoveries and protecting their priority. They were not just concerned that truth extend into the world. They wanted their names attached to the truths. And from my own experience, I know that the laying down of truth confers great unconscious advantages. It helps keep the fear of death at bay.

Yet now looking past all the psychological complexities in its activities, in the end science might be most important not for what it gives to its professional practitioners but for what it gives to everyone. Science produces knowledge about how the universe in all its corners works. It informs everyone's mind about reality. It allows those who drink in its results to share in a cosmic bout of thirst-quenching in which the drinkers partake in universals, and thereby gather them into an extended self. Science is not just a means for some to create temporary immortality projects as articles in technical journals; it is knowledge for all in an extended self of knowing.

Particularly important is science's ability to make visible the many connections that are relatively invisible at various levels of nature and culture. To a large extent science is the practice of revealing relationships.

Terror management theory, to cite an example, searches for the way in which the relatively hidden fear of death links people to their very visible and various behaviors in speech and action. Relationships are nearly always less conspicuous than the things they link. This idea of the invisible linking the visible helps us model the extended self with its complex web of associations between people as individual minds.

One specific metaphor for the extended self comes from ecology, the science of relationship among organisms, and between organisms and their environments. I often experience society as similar to ecology, an ecology of mind, to use the term of Gregory Bateson.[93]

I vividly recall my visit in 1999 to Westminster Abbey, in London. I read before entering that within I would see the tombs of kings and queens and many other notables. I salivated to see the spacious architecture, but as for those decorated tombs and monuments, well, I thought I could miss them. Yet within seconds I forgot about the building and was entranced by the iconic records of past people. I found myself wandering around in a drunken state of "Whoa, whoa, the atmosphere, the ambiance, what is going on here?"

The experience was perhaps closest to the first time many years earlier that I walked into Muir Woods, the magical valley just north of San Francisco. There I beheld the majestic, behemoth trees tightly integrated into the so-called cathedral redwood forest. Just as Muir Woods is imbued with the sense of an extended organism, the Abbey holds the experience of the extended self. Perhaps this experience is one answer to the issues raised by the revelations of terror management theory. If we are bound to defend a worldview and try to elevate self-esteem in our reaction to the awareness of death, what kind of worldview and self should we have?

The cathedral was jammed with visitors. It was Saturday, the day before Easter—bad enough for one who might seek some solitude in the cathedral. Worse still, on the day before, Good Friday, the Abbey had been closed to the public, which created a multitude that rushed to the gates on Saturday. It seemed that half of the more than three million yearly visitors, a number provided by a smiling man in a red robe, were there that day. You had to go with the flow of the packed crowd, like cattle in stockyard chutes. But one can endure being herded along with the crowd, to pay homage to sixteenth-century Queen Elizabeth and other deceased luminaries.

A hand-held audio set worked wonders to help concentrate my mind by providing me with a private tour. And after a while the power of that stone Muir Woods sparked wonder in my brain. I was in heaven, with realizations coming apace from right and left. The sense of history simply seeped out from tombs of ancient kings and queens, as I sensed the intrigue, perseverance, ambition, and pace of their full lives. I also found inspiration in the less-visited corners and spots along walls. Words of wisdom, such as "MANS LIFE IS MEASURED BY THE WORKE, NOT DAYES," are so much more weighty carved in stone than rolled out as mere ink on printed page. On this visit, I found the following along the north aisle of the nave, on a plaque near the body of a Mrs. Mary Beaufoy:

> *Reader, who e'er thou art, let the fright of this Tombe*
> *imprint in thy Mind, that Young and Old (without distinction)*
> *leave this World, and therefore fail not to secure the next.*

What a perfect example of terror management! First comes the terror, "the fright of this Tombe," then relief, "fail not to secure the next

[World]." Such a gem from a past psyche, however, is not the main treat of the Abbey. Neither is the further smashing of my own, pathetic, fame-seeking ego by the numerous achievers of fame in all walks of life, seen in the poet's corner, the statesman's hall, Henry VII's chapel—overall, a world of outstanding people, a world of accomplishments.

In the Abbey it is child's play to perceive the extended self. These dead reach right into me. They are people who have affected my life. Kings, queens, scientists, poets, thinkers, musicians. Some I know well, others not at all. So I am not conscious of how they each have wrought their effects on me. That doesn't matter. They have affected somebody, and these "somebodies" were alive in the world and have affected others who have affected me. Sometimes the most profound effects come from originators you don't even know of. Who influenced the course of the history of music? Or that of the written word? Many in the Abbey and many, many more. What about politics? Political structures ripple down to us from past leaders in such a manner that only the astute experts could say how a twelfth-century decision evolved and thus lives on in the world today.

Now consider the valley of giant redwoods. A single tree is easy to stand before and focus upon, perhaps to study its height or to gain a detailed look at its rust-colored bark. But what about Muir Woods itself? First of all, it must be experienced in time and space, with a hike. Then it can be conceptualized by memory of the hike. But if one tree is over-whelming, how much more so is the woods! The situation is similar in Westminster Abbey. Stand before a monument to one poet, say Shakespeare, and a normal state of awe can be fairly well maintained. But take in the whole Poet's Corner, and one begins to reel from the "woods" of talent and recognition of the outpouring of collective, literary humanity. That's it—the Abbey induces a feeling of something overwhelming and difficult to perceive, because it must be done conceptually with limited brain apparatus, yet it is nonetheless sought because of its gifts that enrich who we are.

The ecological metaphor seems to serve well for the Abbey experience. As one noses into some small corner of an alcove just to delight in what statue or plaque it may hold, ignoring the guidebook, exploring as one might explore nature on a sunny day too glorious to plan each step, one asks: Who is this person in the alcove? I don't even recognize the name, yet the former life is palpable. Similarly, in nature, who is this insect hidden in the folds of the purple thistle flower? I don't know its name. Freely investigating details, in either place, can produce a pleasant buzz from perceiving the sublime complexity of individuals in a larger whole. The organism in nature is like the person in society.

However, this analogy with nature is not yet quite right. The statue in the Abbey is for the dead, whereas the insect in the flower is alive. Let us attempt to forge a more exact analogy by asking, "Where in the ecosystem are the dead?" For that look down to where the worms feed, where roots probe and microbes rule. The soil. Into that dark, moist realm fall the needles, the leaves, the feces, the corpses. Certainly this is the place in nature analogous to what the Abbey represents. In the soil are products from creatures past. Litter in the top zone is just a year or two old. Beneath is former life now aged to decades-old detritus. Further down the time scale of the particles approaches centuries. In the Abbey, we sense the soil that feeds up from the past and into our present human culture. The products of former people are nutrients to us.

Recently, soil ecologists have revealed a new level of connection between many types of trees. The intimate association of trees (via their roots) with the so-called mycorrhizal fungi have been long known. But the dynamics of the association are difficult to study because the fungi are so tiny—a fraction of a human hair in diameter—and therefore are nearly invisible thin threads that radiate in huge numbers outward into the soil from their more visible clumps, upon (and even within) the roots of nearly all species of trees.

The soil ecologists labeled some of the carbon dioxide taken in by a tree's leaves so that they could track its path as it became carbon compounds traveling within the above-ground tree, down through the roots, and even into the threads of the attached fungi. Such transfer of food from tree to fungi had been known. It's the price the tree is willing to pay for the benefit (via the fungi) of increased nutrient uptake of mineral elements from the soil, which the tree also needs. The surprise came when some of the labeled carbon appeared meters away in the soil and even in neighboring trees. The fungal threads must have been acting like highways for the carbon, transporting it as far as trees of different species. Terms like "original world wide web" and "wood wide web" began to make their way into the news about the discovery. Obviously the separate trees were much more intimately connected than anyone had suspected.

I like to think of this discovery as analogous to human society. In the Abbey the image of the extended self is intensified by the overwhelming presence of the influences in so many human endeavors. We all share nutrients, both with each other now and with others from the past.

———

The recycling of the dead in the soil of forests can increase the abundance of trees as well as carbon, globally (with the atmosphere included as part of the system), by two hundred times. How much more amplified is our degree of consciousness because we are social beings who live with others and pass ideas around? Is it two hundred times more consciousness? I would hesitate to put a number on it, but the amplification is obviously great. A child deprived of social contact becomes a wild child, without language, without affective sharing with others. How lucky we are to be living among other people with whom we pass ideas into the psychological nutrient pool of sound and words and print, to ingest them as we need for growth. The soil and atmosphere systems serve as a metaphor for just how much more lively, how much more conscious we are as a social system from the cycling of the small, the nearly invisible, through the globe-circling loops of our communication currents. Words, concepts, emotions, numbers, ideas—we build these like ions into larger structures, most notably the concepts of self, autobiographies in our unconscious, and our abilities to recognize things and love people. We are richer because of the circuits of connections.

With this vision we can better understand Hamlet's despair in one of literature's most famous scenes that involves the contemplation of death. In the graveyard scene with Horatio and the gravedigger, Hamlet contemplates the rotting skull of the court jester Yorick, who entertained Hamlet when he was a child.[94] Where are your gibes, gambols and flashes of merriment, now? Hamlet becomes still more melancholy and eventually questions the end of mighty conqueror Alexander, famous among the famed. Was a smelly skull Alexander's fate, too? Surely, says Horatio.

Then Hamlet takes us through the physical process of Alexander:

Alexander died, Alexander was buried, Alexander returneth
into dust; the dust is earth; of earth we make loam: and
why of that loam whereto he was converted might they not
stop a beer barrel?

Here we see one version of the biogeochemical cycles, how some matter comes around from former life to increase present life. Mighty Alexander to stop a hole! Is that enough? Not for Hamlet.

In our vision, we can see that Hamlet is mistaken in equating Alexander with his matter. Alexander was a person who now affects Hamlet himself, in the present, a personage now dead but within the minds of the living. Alexander in Hamlet is part of the extended self. Hamlet looks only at the clod of dirt. Instead he could have been glorifying in Yorick or Alexander, both still in his mind, or in Horatio still living and also in his mind, all forming within Hamlet an ecology of mind.

But is this enough? Living on as memories in others? Well, while we are alive we also exist as memories in others and in ourselves. So after we die we are solely in others, a severe diminishment of our condition. Of course, the key is how we affect others, not just their memories of us. But we have these effects on others while we live as well, and after we die our influence fades and fades. We do return to lesser units. That's reality. But at least now, when alive, we can concentrate on the psychological circuits. We can enhance our self-esteem by paying attention to our participation in these circuits of psychology, rather than merely the cycles of matter. The main point is that in the cycles of the extended self we participate in amplifying the consciousness of others. I suggest taking advantage of this and trying to experience the extended self as much as possible when we are alive, not waiting until death. After all, the important goal is to evaluate death during life, and have the evaluation affect life positively.

We can lament the dehydrated beetle, the fall of a gnarled tree, the deer killed by a mountain lion. Yet we know that other beetles, trees, and deer will rise again, similar to those that died. This arising depends on the cycles of matter, turned into life by life. (Extinctions are thus particularly tragic because the species is ended.) So, too, we can learn to see our psychological selves. Let ecology serve as a metaphor for our thoughts, for our unconscious selves that have memories bound into ideas, emotions, and shared activities with others. Beetle, tree, and deer—these live connected in webs of amplification for all life. They die. But while alive they thrive in the interconnected dignity of the circulating lives and deaths of others. When they die they stay in the web. They were each unique, vibrant entities. When they die they have dignity because of their connection to the whole.

So, too, with those in the psychological web of amplification of consciousness. Each of us contributes. And each of us should be thankful for all those in the past, ever newly circulating and forming our world. Each cannot be an Alexander in conquest, a Darwin in science, or a Buddha in religion. To use an analogy from the ocean, most of us are not whales, but plankton. Yet each plankton cell is an integrity of being. It both takes from the whole and gives to the whole. Death is its final giving. And our death, taken at the psychological level, is a kind of final giving, because others will awaken more to their own death and at least somewhat alter how they orient themselves to that future event.

If we follow the monist idea, then death is the end of the individual self. But what about the webs of influence, the webs of help, even the webs of competition? These interconnections certainly run deeper than we ordinarily sense, because so much of our language conditions us to feel separate. Such habits of psychological separation are likely related to our instinct to preserve our unique, living, fleshy body. As we change how we consider the self, however, we change the perception of death. The more we realize that we are interpenetrating ripples on a river of being, the less we fear individual demise.

Thinking about our connections to others gives us a concept of death with interconnected dignity. We can learn to pause in our daily, busy lives and see these psychological circuits as nutrients from which we are built and to which we also contribute. Do not wait for funerals to build the experience of the extended self. Toward the psychological web have awe—and gratitude. Be thankful for it and for the ability to have a wondrous vision of it. Death then is less troubling. The moving finger writes, and then it stops that line. But have gratitude that the finger moves on to write another line, for another person, not so different from me, who will also say "me."

If death produces fear that turns us into creatures over-eager to defend worldviews, let us be conscious of which worldviews we choose. A worldview that shows us as extended throughout the psychological world of humanity, as the trees' roots and fungi are extended throughout the soil, changes us as we live within that view. I am grateful for the years of being an individual sprung from, and contributing my roots and leaves to, the social soil, grateful for my effect on others and for theirs on me. We exist in an ecology of mind; physical brains are molded by their development within the soil, creating pathways of experience that are tied together. When one brain blinks out, so flattens the complex ripple of mind born from the workings of that particular brain. To fulfill our desires to live on after death, we need only probe more deeply into what we actually are as individual selves. We will find that we are not encased in skulls. We are extended out—indeed substantially merged with others. What a worthy worldview to live by.

PART THREE

biosphere

SEX AND
CATASTROPHIC
SENESCENCE

I recently witnessed a mass death among male ants. In mid-July, with the late sun sunken far enough to already throw the valley into shade, I was walking to the edge of a small cliff to sit and think. On the way my eye was drawn to a spot on the path that held a remarkable phenomenon. Hundreds of tiny ants were swarming around the hole that led into their underground nest. Within the multitude were here and there a few that sported much larger bodies. And they had wings. They were males, readying for nuptial flights. The monsoon rains had begun about a week earlier, and in this dry country many species of ants time the maturation of their males for the wet season and the abundance of food the coming weeks would bring.

Compared to the frantic chaos of the smaller workers, these males behaved very differently—more deliberately. They would carefully emerge from the hole, move around a bit as if surveying the overall situation, then pop back underground, only to emerge again. During this routine the work-ers would swarm right over the seemingly more sluggish males. Finally, I watched rapt as a few males crawled away from the horde, one to a twig nearby, another to a grain of gravel, and then like helicopters, they spread wings and launched skyward. Several slowly wended upward for a few meters, circled in the air and then disappeared—too small for me to see. Several more departed, then still more.

Energized by the spectacle, I returned to the trailer for a mat to spread outside on the dirt driveway, planning to do some stretches as evening unfolded. I had no sooner begun when male ants began landing on me and the mat. There must have been nests all over that were busting out with flying males. Surely there were females in the sky as well, but it was the males who overwhelmed the air. Their black bodies contrasted against my yellow teeshirt, so as they landed on me I could look closely. They had such incredibly tiny heads. I had never seen such a disproportionate head on an ant before. Then it struck me: With such pinhead heads, they were obviously not going to be feeding.

Not only were they not feeding, they were dying. All around me they showed signs of trouble: hobbling like tiny drunks, stopping, staring. Dying seemed to take a good half hour. Near their demise, they just stood still. Some tiredly swept their legs and antennae in the space around them, but only for short intervals and obviously with failing strength. Some attempted additional flights, but usually crash-landed, often upside down just a few inches away from where they took off. I picked one up by its wings. They were spread, but the ant's legs were all crumpled. The creature hardly reacted to my touch. In the fading light it was dying in my hand. I had a magnifying lens with me and I reveled in its blue and purple, intricate and iridescent wings.

I decided to save a few for examination later. Several ants just clung to my shirt, nearly lifeless, as I walked to the porch and flicked them into a container.

The next morning I took out a microscope to examine them. Their abdomens were relatively huge, probably for reproduction. And, surprisingly, the ants were still not completely dead. This fact I could discern only under the microscope, because the genital tips of their abdomens twitched ever so slightly. That was spooky. Their thoraxes were large, too, obviously endowed with muscles, the bulk of which were visible as powerful attachments leading into the wings. And the heads! They were indeed exceedingly tiny. I could see their beautiful eyes with about two hundred facets each. The mouth parts appeared degenerate and nonfunctional, lighter in color than the dark head and looking like short, furry, weak antenna. There was no sign of any mouth opening or jaws, or of any mechanism for feeding. Later I was told by an expert that the males do have mouth openings through which, during development, they were spoon-fed, as it were, by the worker ants. But lacking biting components, males are unable to forage for themselves. After their nuptial flights there is no one around to feed them anymore, so they die.

Whoa, I thought. Here is a creature that was designed to die. I had read about the phenomenon in what are perhaps the most famously ephemeral insects, the mayflies. Adults of the more than two hundred genera of order Ephemeroptera emerge from their final underwater nymph stages into delicately winged forms ready for sex but not for eating. Like the male ants I saw, the mouths of mayflies are degenerate. It is as if evolution had designed them to be doomed.

Death in these dramatic cases appears to be programmed. Creatures who can't eat? Creatures who will die soon after mating or after attempting to mate? Mayfly nymphs crawl around underwater as voracious predators with strong jaws and very viable digestive systems. Their genes, which control their metamorphosis into the flight-ready adults, seem to have neglected to provide them with mouths. To add insult to injury, their adult abdomens could not digest any food anyway and are usually inflated with air. Obviously the concerns in the design of the mayfly adult are flight and sex, not survival. These adults might only live minutes, or hours. If we feel at all sorry for their short adult phases, we should take note of their full aqueous lives, with often ten or more nymph stages, typically spread over several years.

Clearly genes are also involved, somehow, in the demise of organisms not so dramatically handicapped. Such as us. We senesce, even in the best circumstances. Senescence is not a disease. It makes us more susceptible to disease but it is, unlike disease, part of our genetic program. This is why most humans age and fall apart at more or less the same rate.

Aging, which leads to death, is species specific. Humans generally live longer than dogs, which live longer than mice. The lifespan of mice, typically several years, beats the several-month lifespan of the tiny fruit fly. Even smaller, a sixteenth of an inch long, the soil roundworm used in genetics labs worldwide has a lifetime of only several weeks in normal growing conditions. These durations are set within the genetically determined metabolisms of creatures.

Clearly there is difference in our slow, senescent breakdown compared with the forced starvation and rapid decline of the male ants, mayflies, and various species of butterflies and moths. But some of the general principles of how death is related to the evolved lives of creatures are common to all. Let us ask the question about the design of death for the dramatic cases in which death follows on the heels of sex, like the male ants or mayflies. This is a pattern that

Caleb (Tuck) Finch, a well-known biologist of aging, calls "rapid senescence."[95] Its rapidity is catastrophic. So we might call it "catastrophic senescence."

Why should genetically determined metabolisms cause death? Considering the presumed genes of such metabolisms, we should expect to be able to justify their presence by some benefits that they provide for themselves via the organisms that carry them within the context of populations within a particular species.

It is difficult to understand how what seems to be programmed death would help the genes of an organism, because death removes that organism from the competition to spread its genes during reproduction. The male ants who die are now out of the picture. But perhaps their death benefits the next generation. Could it help the young that the males die and thus do not return to the hive to compete for resources? In general, do creatures such as we age to make room for the next generation? This sounds like a viable idea, a sort of sacrifice of one generation for the next.

But this logic is absolutely wrong if we consider the way that genes are stabilized in populations during evolution. This conclusion can be reached as follows.

Assume we have some male ants with mutations that give them mouths and jaws. They can mate and then—unlike the real ants I saw—feed themselves and continue to live. The death genes have thus been overcome. Furthermore assume these mutant males do not go back to the hive but hide out in nature. They stay by themselves for a year, feeding, biding their time, perhaps going dormant through the winter and until the next mating season. In this state they are not directly competing against the others in the hive. Then at the right time next year they emerge from hiding, ready for an additional round of mating. With the mutation of mouths the entire package of genes in an organism has a better chance to be spread to the next generation. With extra chances of mating, the genes of these mutants, all else equal, would soon dominate the population. "All else equal" means that no additional resources had to be given the mutant males when they were originally raised as larvae in the hive. "All else equal" also means these mutant males had at least the same capabilities of securing females as other males in the hive during their mating flights the first season. The fact that additional mating flights are obtained from the same organism in subsequent years is what gives these new mutant genes the selective advantage over time.

The crucial phrase here is "all else equal." But for the moment, let's summarize: From the viewpoint of the genes within a particular individual, if all else is equal, it always helps to maximize the spread of those genes to have a situation in which the individual lives longer and thus reproduces more often. Therefore, long lives should be selected for individuals within a species during evolution, all else equal.

But long-lived, mutant male ants did not evolve, at least not in the species I saw. And longer-lived, feeding adults of mayflies did not evolve. And male tarantula spiders who lived past a first season of mating did not evolve. And fruit flies who live for years instead of months did not evolve. And humans who live and are reproductively fertile for a thousand years did not evolve. Because death is such a disadvantage to the furtherance of the genes carried by the organism that dies, the apparent programming of death into the biological structure of creatures must mean that all else is not equal. Specifically, we need to consider the following rule of evolution:

When disadvantages arise in evolution, they do so as a side effect of changes having overwhelming benefits overall. [96]

Death due to breakdown of the body halts further reproduction and is thus disadvantageous to the propagation of the creature's genes. So the fact of senescence implies that some characteristics of the organism that are associated with this process toward death confer advantages (in gene propagation) that outweigh the disadvantages. Say, for example, that some of the suicidal creatures have mutations that eliminate or at least inactivate the death genes. Those creatures will live longer, reproduce more young, and these pro-longevity, anti-death genes will spread and soon dominate the population. Via evolution, it does not seem to make sense to have genes that are exclusively aimed to shut the creature down. Lack of mouth parts during development is related to genes, to be sure. But there must be more going on.

These creatures with catastrophic senescence inspire the ideal forum in which to think about these issues. In the science of aging, the idea that senescence is an unfortunate byproduct of some overwhelming benefits is known as the "disposal soma" theory.[97] Soma refers to body, so we might call the idea the disposal body theory. The underlying concept revolves around the passive sacrifice of the body in the pursuit of successful reproduction. According to the disposal body theory, after reproduction there is less and less selection pressure to keep the organism healthy. Aging is a byproduct of this lack of selection pressure. Deleterious genes whose effects come later in life, after reproduction, can weaken the body and yet the genes remain. Also, genes that specifically aid efforts in reproduction, but which also create havoc later, can remain stable in the organism during evolution.

I say "passive" rather than "active" sacrifice to emphasize how the demise of the body is a byproduct in the disposal body theory. A good place to look at how the passive sacrifice works is in what many would call the most amazing example of catastrophic senescence. It involves a big, charismatic fish, the Pacific salmon.

All seven species of this salmon exhibit catastrophic senescence. They hatch from eggs in an upland stream and spend some time migrating downstream while feeding. They reach the ocean and spend, typically, two or three years out at sea, growing large. When reproductively mature, they migrate back up the stream of their birth to spawn. The journey is a grueling, grinding haul up rocky rapids, in a mass migration. Some turn color from brown to red. The male grows huge jaws for fighting with other males over the privilege of mating. The female deposits eggs in a streambed hollow she has formed with her tail or flippers, while the male fertilizes the eggs with sperm squirted on top of the eggs. Matings continue. Then females and males die.

Measurements on migrating, spawning salmon have shown elevated blood levels of a steroid and adrenal hormone called hydrocortisone.[98] These levels damage the fish tissues in ways that resemble Cushing's disease in humans, an over-active adrenal syndrome. As a result, a salmon's liver, spleen, and kidneys degenerate. The fishes also stop eating, which wastes their bodies. Furthermore, salmon undergo chemical stress in the switch from marine salt water, where they lived for several years, to the freshwater stream for spawning. Finally, the stream migration itself is an ordeal, with abrasions against rocks and violent leaps to gain purchase toward the upland waters. Elevated steroids may be necessary to push forward the heroic feat. Thus a physiological behavior critical for species survival also serves as executioner.

The Pacific salmon show that it is sometimes worthwhile to have the body crash as a result of putting resources into reproduction. But after all those years of feeding and survival in the ocean to build a large, powerful fish, why haven't the biological tradeoffs been worthwhile to evolve a salmon strong enough to survive a journey back downstream to feed in the ocean for another year and then upstream again the year after for more reproduction?

The answer is not known. But biologists would generally agree that in the Pacific salmon we have an example of the body being disposed of in the furtherance of mating. The death is a passive sacrifice because it is a byproduct of what is required for one successful mating run. Death results not as a primary effect but as an unfortunate spinoff from the real vital attribute: a bold and successful mating run. The hydrocortisone probably helps the fish in its ordeal, even though this internally generated drug aids in killing it later.

The idea is this: If a physical or biological characteristic of an organism helps it reproduce and spread its genes to the next generation, and if that same characteristic inadvertently slides the organism downhill toward death, the characteristic can evolve and be stable in the population if the total increase in offspring now outweighs the decrease in later offspring caused by the death. In the disposal soma theory, we must ascertain how a characteristic increases reproduction averaged over a shorter period of time with quicker death, compared to a lower level of reproduction averaged over a longer time without death. In numbers, a thousand offspring in one year of mating will outweigh two hundred and fifty offspring per year over two years of mating, which totals only five hundred offspring. What the Pacific salmon gained by evolving a way of reproducing in a single mating season must have outweighed an alternative lifestyle in which several years of mating yielded fewer baby salmon for the overall total.

It boils down to a question: Which provides the most spreading of genes—maintenance of the body for future reproduction or disposal of the body for reproduction now? Catastrophic senescence is one pattern that answers this question. We have seen this pattern in some insects and fish. There are even a few examples in mammals.

In Australia live tiny creatures sometimes called marsupial mice. Their ten species show that even in mammals, the pattern of sex and quick death can be a reality. Males are the ones who die. The cause is probably stress. In the lab colonies, males just under a year old undertake repeated sprees of copulations that can each last up to twelve hours. Then they die. Their deaths resemble that of the salmon. In the month before, the tiny marsupials' adrenals increase in size and the steroids in their blood rise five-fold. Other hormones elevate as well. If the males are captured before mating, and prevented from sexually dissipating themselves, they will live to three years, a two hundred percent increase in lifespan. Females survive for an additional year after their first litter, but the odds of their having a second are very low.[99]

We can now understand the rapid demise of the male ants. It was not a matter of all things equal, meaning mutant ants could have simply evolved with mouth parts and an ability to live longer for additional mating. Things would not have been equal. It would have cost energy to build massive mouth parts. Assuming that the developing male larvae were all fed the same amounts by the worker ants in the hive, having mutants grow mouths capable of self-feeding would have meant some other part of their body would have been smaller. Weaker wings would have been one possibility. This might have taken them out of the first-year competition altogether and thus it would have been of no benefit to continue to live. Genes that help an organism live long but hinder it from reproducing will not carry on in the evolutionary game. Pledge all resources toward continuing the life of an organism, and it never reproduces.

Nature was not cruel in giving birth to adult mayflies with dysfunctional guts and no mouths. Nature saved resources for better wings, or more sperm and eggs, or more gyrating nuptial flights. We don't have the exact answers. The experiments have not been done to enable us to know precisely what body part was sacrificed as unnecessary so some other body part could be built that rendered better overall success in reproduction. But the logic has been laid out in evolutionary theory. Trade-offs have been made.

And think of the mutant ant making it through the winter. It probably would not just be a matter of feeding and therefore nutritional self-maintenance. What if its wings became tattered? To replace wings (which no insect can do) would require new metabolisms. And in living longer it would be more prone to disease. Fighting disease might require an improved immune system. In other words, living longer requires increased metabolic complexity. Long life comes at a price. That is why we cannot just consider a redesigned animal for longer life and keep all things equal. Sure, with all else equal it would be good to put off death and keep the creature around for more reproduction. But keeping the creature around would entail a different kind of creature and thus complexity, more genes or more complex enzymes present all the time. And that could diminish success in competition for reproduction.[100]

Common patterns can be invented independently by different organisms during the process of evolution. It is obvious, for example, that the pattern of catastrophic senescence in both mayflies and Pacific salmon is not shared because these creatures were derived from a common ancestor. No birds yet have been found with catastrophic senescence. But seeking the general concept, we might turn to a different type of organism entirely. What about plants?

Catastrophic senescence abounds in that kingdom. Gardeners call them annuals, which degenerate relatively rapidly, subsequent to reproduction. For many meadow flowers the whole organism dies after the seed is set, grown, and sufficiently matured for dispersal to take place successfully. Even though all this structure of roots, stalk, leaves, and flowers was built over the course of a growing season, somehow it is worthwhile to let the whole biological, biochemical shebang die after going to seed. Evolving these plants— tens of thousands of species, worldwide—to be viable longer would have presumably taken energy away from their first year's reproduction and thus made them, for their ecological circumstances, less competitive.

The catastrophic senescence following seed release can also occur in plants that have grown for much more than a year. Some of the most famous examples include the cacti called century plants. They grow in rosettes of tough, pointy, fat, stalk-like leaves. After many years, a fantastic stalk, sometimes twelve feet high, rockets up and spreads out an inverted candelabrum of flowers. Then following fertilization and seed growth, the entire plant dies. The number of years the rosette takes to grow prior to this crucial event can be many fewer than the proverbial century, depending on the species. But even ten or twenty years is an impressive interval for the slow but sure growth, all for one glorious round of seed production.

Bamboo is another spectacular example. A Chinese mainland species of the genus Phyllostachys will grow for more than a century before flowering, seeding, and then dying. Incredibly this bamboo, as well as many other species with shorter cycles, will flower simultaneously with all the plants in a geographical region, and thus die in what seems to be a mass suicide of the bamboo cult.[101] This behavior is reminiscent of the cicadas. These insects, like the bamboo, have a slow buildup toward adulthood of many years in a juvenile phase. When mature, the cicadas emerge from the ground en masse, to fill the night with staccato chattering during their mating rituals, before dying.

These dramatic examples show us how the plant's lifestyle is oriented toward producing seed for the next generation. The plant must be healthy until that point. Afterwards, theoretically it can die and the species will continue. This conceptual logic becomes actuality in many species. Are the stalks of annual grasses merely supports for the seed head? Are not the leaves of globe mallow the caregivers for its orange flowers, and these in turn the lures for pollinators to start the seeds that close the plant's life?

Catastrophic senescence is a dramatic example of "death, thus life." Death does not occur to make room for the next generation, but as a result of this generation's effort to propel the next one into the future.

LIFESTYLE
AND LIFE SPAN

Amazingly, some creatures don't appear to age. They grow chrono-logically older, of course, and they do die from predation, disease, or overall wear and tear. But their susceptibility to death—by being more prone to disease or less able to fight predators, for instance—does not seem to increase with age. If such susceptibility is the hallmark of senescence, these creatures do not seem to senesce. Tuck Finch, in characterizing this life span pattern, has termed it "negligible senescence."

Two of Finch's prime examples of negligible senescence are a species of tortoise and the quahog clam. The tortoise has a recorded life span as long as 150 years. The ocean quahog, a bivalve mollusk, has reached 220 years.[102] Other candidates for the phenomenon include several deep sea fish. Individuals of the Northwest Pacific's rockfish and the Southern Hemisphere's orange roughy have been logged at around 150 years.

The point is not that these creatures live so much longer than humans—they might or might not. The issue is that in these species the individuals who are longest-lived (so far as we know) do not degrade from internal causes that would make their chances of death increase as they age. Some species with negligible senescence maintain high reproductive output despite their increasing years. In fact, as in the case of the lobster (another probable species with negligible senescence), egg laying can become more copious with age.

Among the oldest known trees are the bristlecone pines. They are the Methuselahs of the American West, particularly in California and Nevada. One individual has been dated at 4866 years (as of 2002), but some are probably over five thousand years old.[103] These pine trees, from a variety of ages between seven hundred years and nearly five thousand years, have been tested for pollen germination rates, seed weights, and seedling growth rates. Intriguingly, the seeds put out by the oldest bristlecone trees were as high in weight, germination, and subsequent growth rate as seeds from the younger trees.

Curiously, the trees at the highest elevations reach the oldest ages. One might think that these altitudes would limit the trees' lives because of the harsh weather conditions. But instead, in such lofty air the trees reach their full potential in age (but not weight), probably because they are less exposed to insects and likely competitive tree species.

Finally, with regard to plant models that seem to defy the normal pattern of senescence, clonal plants deserve a mention. Clonal species, such as aspen trees, can have their aerial parts die, but they remain alive as a rooted individual. From this biological center in the soil, new ranks of shoots can be launched, often over a wide area. Thus a cluster of aspen trees is usually only one tree.

The creosote bush, in the American Southwest, spreads in this clonal manner, with new growths occurring in an ever-expanding ring. The older growths die and rot away from the center. Age can be estimated by the diameter of the circle of a creosote plant. Some rings have been located that might be nearly twelve thousand years old.[104] Not bad, for a bush.

The state of Michigan has a champion of size. A single individual fungus has distributed its gray threads underground over nearly forty acres; it weighs more than a thousand tons, and is one thousand five hundred years old.[105]

Let's be clear. These patterns we are discussing are the natural, upper-end periods of life span under ideal conditions for organisms of a particular species. This is different from life expectancy following birth of all individuals. For example, a female salmon might lay many thousands of eggs. Most hatchlings will die in their birth river before reaching the ocean. More will die during the two or three years in the ocean. If after birth the ideal salmon has a two-year interval (according to that species, say) in the ocean before returning to the birth stream to spawn and then die, we would say that the life span is two years. But that is the ideal condition. The average life expectancy of a hatchling might only be measured in weeks. The salmon is subject to a high death rate due to environmental factors, including predators. This high mortality rate, as we will see shortly, is crucial for understanding the natural life span, which so definitely differs among species.

_____

We have said that the seemingly strange cases of catastrophic senescence can be understood as some sort of trade-off in which the advantages of a single burst of reproductive effort and the sudden demise of the individual outweigh what would be lesser reproduction even over the longer run with a longer lived individual. But how do we explain the spectrum of types of senescence? We must look more deeply into the nature of the trade-offs played in the game of reproduction via the coordinated lives and deaths of creatures of various species.

To continue living, creatures must endure a continuous onslaught of biological and physical hazards. The situation reminds me of that peculiar brand of auto sport called the demolition derby. Drivers who choose to enter the derby roar around an arena with the purpose of smashing all other cars to put them out of action. When one car alone still runs, that one is declared the winner.

Organisms do not have a choice whether or not to be in a derby. They are in the game of avoiding being eaten and of eating (or, in the case of plants, obtaining nutrients). To draw out the analogy between the demolition derby and creatures in ecosystems, let us assume a hypothetical situation in which all cars are in a derby.

Suppose that by law, five years after you purchase a new car you are required to be in a derby for all cars that age, in groups of one hundred. The last car rolling is declared the winner. If it is you who wins, your car (and your car alone) will be repaired according to any damage received in the derby and guaranteed for five more years. But this repair and guarantee will be provided only if when you originally purchased the car you had the foresight to buy a derby maintenance and repair option, a kind of special derby insurance.

Would you buy the option? It depends, of course, on how much it costs. You only have a one percent chance of surviving the derby. For a few dollars extra at the time of purchase you would probably pay, because for that little extra expense you would gain a lot, should you make it through the derby. But if, for example, the option cost you two thousand dollars extra on a twenty-thousand-dollar car, buying the option would be throwing your money away, statistically speaking. You would be paying ten percent extra that would help you only if the one percent chance of surviving the derby comes true. That would make no sense.

In the analogy to life, the derby represents the external causes of death to organisms from predation or general physical hazards in the environment. Making it through the derby is what the creatures must do virtually every day of their lives. What are the odds of surviving the predation derby in nature? The higher the odds, the better the gamble it is for the creature to have purchased the maintenance and repair option. What is that? The biological option is exactly maintenance and repair, and it comes from internal designation of resources to genes and their outputs of enzymes and other molecules that keep the creature healthy and, after maturity, capable of reproduction.

For organisms, the odds of making it through the derby of predation, starvation, disease, and other forms of externally caused death are not necessarily one percent at the end of five years, as in our car example (for many species they are much worse). But the principle is the same: whether or not the maintenance and repair option is worth purchasing depends on the odds of successfully making it through the derby. If the odds of success are ninety-nine percent, a much higher price would be worth paying for maintenance and repair during the preceding years than if the odds are only one percent, as in the example. For organisms, when they have excellent odds of continuing through the gauntlet of predation, it makes sense for evolution to have given them intrinsic longevity by committing energy resources to their metabolic maintenance and repair. Therefore, for organisms that have a good chance of surviving the many demolition derbies that each must enter, it will be worthwhile to have evolved intrinsically long lives.

How in nature do creatures increase their odds of surviving? One way is sheer size, as with elephants, for example, and hippopotamuses, creatures tough for predators to tackle, and whose hugeness may discourage any attack in the first place. The rule of body weight bears on this advantage.[106] Humans live longer than dogs, which live longer than mice. The life span of mice, typically several years, beats the several-month life span of the tiny fruit fly, which bests the soil nematode, a sixteenth of an inch long or so and fated to spend only several weeks on Earth given normal conditions. There are exceptions. A lion outlives an African buffalo, though the buffalo would win in a tug-of-war. Size, however, should best be considered as relative within a group of like types. Large salamanders, for example, are not nearly as large as medium-sized mammals, but they are large for salamanders, and they do have a relatively long lifetime of decades.[107] Even with the exceptions, statistics show the rule generally holds.

In contrast, small creatures suffer high rates of predation. Think of tiny, flying insects being snagged by dragonflies, robber flies, or bats. For many insects, because the odds of making it through the ongoing demolition derby of predation (or other environmental hazards) are so low, there are no reproductive payoffs for them to have evolved more diversion of their bodies' resources into the extra maintenance required for enhanced longevity. This might be why catastrophic senescence following mating is relatively common in insects.

The fact that an organism's ability to maintain and repair itself is related to its death rate—due to its position in ecology (which determines the hazards of its specific demolition derby)—allows us to start analyzing some unresolved issues. For example, why are some creatures exceptions to the average trend of life span increasing as a function of weight?

Of all the orders of mammals, bats have the highest percent deviation above the mammal correlation of size versus longevity. On average across a range of body weights, they live about three times longer than other mammals of the same weight. Bio-gerontologists Steven Austad and Kathleen Fischer hypothesize that the longevity of bats derives from their lifestyles.[108] Their ability to fly makes them much less vulnerable to predators than are ground mammals. Also, bats are out primarily at night and therefore are difficult for many potential predators to see and catch. They usually roost in large groups, which offers protection. And consider those inaccessible roosting sites. Hanging hidden up under rock ceilings, for instance, does not exactly make them sitting ducks (or bats). As a result of these superb defenses, their metabolisms, honed by evolution, have responded by giving them long life spans.

Austad and Fischer further reason that any mammal that can sail between trees should be better than ground dwellers at avoiding predation. They surveyed data for gliding species of mammals: three squirrels, five marsupials, and one flying "lemur." Taken together, these species have life spans that average 1.7 times the mammalian average for their weights. In another of their studies, all marsupial mammals were lumped into two groups: tree-dwelling versus ground-dwelling (species using both habitats were ignored). For comparable weights, the intrinsic life spans of arboreal species beat terrestrial ones by nearly sixty percent.

Possessing a hard outer shell also deters many predators. This is probably the reason that some species of tortoise and clam are among the longest-lived animals known.

Tuck Finch devised a useful term for the wear and tear on an organism during its life: mechanical senescence. Adult insects, for example, cannot replace their cells (except, perhaps, for some in their gut lining and in a few other organs), making irreversible the splitting, chafing, and tearing of wings. So if a fly cannot replace tattered wings, and thus not make it through the derby, there is little need for it to possess repair capabilities for its other parts, such as its internal organs.

Lobsters, on the other hand, can periodically molt, thereby replacing their external, hard tissues. This allows lobsters to have what Finch likes to call an "escape from mechanical senescence."[109] Molting makes it difficult to determine the ages of lobsters (not that they need to know for birthday parties), but estimates have them reaching fifty to one hundred years. Furthermore, these elders show no signs of senescence. Females, as they age, continue to produce eggs with more and more fecundity the bigger they grow. Tumors are rare, and, as a more important characteristic to qualify the lobster as a candidate for negligible senescence, lobster tumors do not show more frequency with age. The lobster's escape from mechanical senescence allows it safely through the derby of death that insects are subject to with their irreplaceable external parts. Thus it has been worthwhile for lobsters to have evolved internal metabolisms that allow very long lives.

Birds are good cases to look at because not only are they wonderfully adept, intelligent, colorful, and overall delightful critters to watch and befriend, they have been studied quite a bit on a number of technical levels. Like mammals, birds of species with greater average body weight live longer than those of smaller body weight, and some live longer than mammals of comparable weight. A typical mouse of twenty grams, for instance, lives about three years, while a canary lives for twenty years, almost seven times as long. Data from wild birds show they live just under twice as long as same-weight mammals in captivity.[110] Of course, it is difficult to find the longest-lived birds of any given species in the wild, so another comparison would be between captive birds and captive mammals. In this case the birds live about three times as long as the mammals, again, of the same weight. Some of the numbers are extraordinary. Scarlet macaws have lived more than ninety years, which is about four times the prediction of that of average, similar-sized birds, and twelve times the mammal average at the same weight.

In a survey of bird life spans, Austad and his colleague Donna Holmes have concluded that the longevity of birds, compared to mammals, is due to the same reason as that for the relative longevity of bats and other flying mammals. They use the phrase "fly now, die later."[111] Because wings give birds the keen ability to escape predators and fly away from other environmental hazards, such as fires or small regions of drought, it has been worthwhile for birds to have evolved a healthy dose of maintenance and repair for their metabolisms, thus slowing senescence and living long.

Death rates make life into what it is, all the way down to the molecular genetics within the cells of a creature. A key concept in the theory of aging at the level of molecular biology is that, everything else being equal, metabolic activity tends to degrade cells. This is because the byproducts of metabolism, such as free radicals, cause mutations or create other damage that cannot always be repaired within cells. Thus metabolism in just the course of daily life gradually damages the ability of cells to both repair themselves and replicate. [112] This might contribute to the fact that small mammals have shorter lives than large mammals. It is known, for example, that the metabolic rate per unit of mass (often measured by the amount of oxygen used per gram of body weight per hour) is less for large mammals and more for small ones.[113] Thus the tissues of large mammals, subject to a lower metabolic rate, should accumulate damage more slowly, contributing to slower aging.

The lower metabolism of large mammals at least in part derives from their larger size and smaller surface area per unit of volume. This means they don't have to burn as much to keep up their high body temperatures. Now as we have seen, large size contributes to longer life spans. So the lower metabolism could also be a way of achieving those longer life spans and not just an aspect of a reduced need for heat production.

Turning again to bats, we might ponder how bats gain their longer life spans, relative to other mammals of the same weight. Could bats reduce their cell damage through lower metabolism, relative to other, similar-weighted mammals?

Austad and Fischer looked into this issue. Just the opposite occurs. Incredibly, bats expend about twice the numbers of calories per gram of body tissue in their lifetime than does the average mammal. So one cannot argue that the longer lifetime of bats is derived from their slower metabolism. Indeed, it is likely that the bats counter the metabolic frenzy within their cells, with its presumed abundant output of waste toxins and waste-derived oxidizers, by bettering the usual mammalian standard for cell maintenance and repair enzymes.

For bats the facts at the cellular level remain unknown. But more insights have been gained for birds. Like bats, birds have high metabolisms, about two to two-and-a-half times that of mammals for comparable body weights.[114] Based on the idea that cellular metabolism is inherently damaging, this high output, all else equal, would kill birds more quickly than weight-equivalent mammals. But is all else equal? The situation becomes even more perplexing when we consider not just the factor of two or more in metabolic rate but the longer life span of birds over mammals. Multiply the metabolic rate by life span and you obtain a number that is the lifetime energy consumption per gram of body tissue for an animal. This number, which reflects internal assaults on the cells, is twenty times higher for a canary than a mouse.

Furthermore, birds have other potential problems. Compared to mammals, birds possess elevated blood sugar levels and higher body temperatures. Both of these factors should also contribute to rapid aging. Sugars have negative effects. They react with proteins to form what are known as AGEs, or advanced glycation end-products. By cross-linking with proteins, it is certain that the AGEs, to put it crudely, basically gum up the interior works of bodies. And high temperatures should promote the formation of AGEs as well as promote all the various types of damage that derive from the active cellular machinery.

We might postulate that for their long life spans birds must counter these metabolic challenges with specific solutions, and there is some evidence that birds generate relatively low amounts of damaging substances, such as senescence-creating free radicals, within their cells. In addition, the evidence suggests that the metabolic machinery of birds might be extraordinarily capable of self-repair. Experiments were conducted on liver cells from mice and three species of birds: parakeets, starlings, and canaries.[115] All cells were subjected to a variety of stresses: ninety-five percent oxygen, which would cause a large amount of free radicals to form in the cells;

hydrogen peroxide, a toxin produced by metabolism; paraquat, a chemical toxin; and gamma radiation, which would cause genetic mutation as well as cell damage. All these stresses caused cell populations to die off, but the cells from birds lasted much longer. The conclusion is that birds do contain better molecular defense mechanisms than mammals, at least as represented by the mouse.[116]

Many details still need to be filled in. But already we can answer the question "Why can birds live so long?" with some positive statements about enzyme repair mechanisms deep inside their cells. If we keep pressing with the word why, we can ask, "Why do they have these repair mechanisms?" Metabolism can ultimately be understood only by considering evolution and its active selection pressure over countless generations to create those mechanisms. To understand the life span of birds, we must let our attention sweep in scale from a bird's cells out to the whole bird; indeed, even to its relationship with its environment.

Stated with the highest degree of generality, birds live longer than mammals of the same weight because their airborne and arboreal lifestyles have made it economically feasible for them to develop cellular mechanisms that enable them to live longer. The answer to the why cannot ultimately be found inside their bodies; it is found in the fact that they are predation escape artists. Their relationship to sky and tree, where they wend their way with artistry, provides the causal impetus for the creation inside their bodies of certain molecular dynamics, including special genetics. Thus, in the evolutionary theory of aging, the micro-level is controlled by, and in some sense is a slave to, the macro-level of what the organism's life habits are, how it avoids death in relationship to the environment and to other organisms.

———————————

To understand the human life span, then, we must look beyond the internal details of, say, why Mr. So-and-so died at a ripe old age, to why he grew old in the first place.

Although we are not one of the special creatures with negligible senescence, we should not have any species envy when it comes to longevity. Like the lion, we are one of the exceptional cases. Compared to what would be predicted for mammals of our weight, humans live about four times longer. Some species of African antelopes, for example, are about our weight, and as average representatives for that weight class, live only about twenty years, max.

Could our longevity have something to do with the evolutionary process that turned apes into humans over millions of years? Thinking along lines of the wings of bats and birds, or the shells of clams, we should search for some special characteristic of humans that could have given our ancestors special skills in surviving the demolition derbies, thus allowing the evolution of a long intrinsic life span to take advantage of that survival rate. Could it be our brains, our smarts? Or perhaps (before the first big expansion of the brain in evolution) our ability to stand upright and run long distances during hunts? Factor in our capability of free hands that can hold stone tools to efficiently butcher game and even dig tubers. What happened over several million years that might have led to longer and longer life spans?

Compared to what? We should look at not only the mammalian average for our weight but the primate average. After all, primates, from small monkeys to big apes, have relatively large brains. How long do primates live? As in groups such as birds or mammals, the subgroup of primates, considered as a set of data, shows a rough correlation between body weight and maximum life span.[117] Those primates that weigh about a pound can live about as long as ten years. Tipping the scales at around ten pounds provides a life span potential of about twenty years. There is quite a bit of scatter, to be sure. But the correlation is good enough that a line through the various data points on a graph of maximum life versus body weight would predict that primates who weigh about a hundred to two hundred pounds should have a life span of about forty years. Some of our relatives weigh about that much. The chimpanzee is somewhat less than a hundred pounds, and the gorilla is somewhat more than two hundred pounds. On the primate line of life span versus weight, they both, in fact, fall very nearly on that forty-year prediction, with the lighter-weight chimp living a bit longer than the heavy-weight gorilla. At the same time, the average primate lives about eighty percent longer than the average mammal.

What about humans? In weight, on the average, we come in between the chimp and the gorilla. Thus our predicted life span, from the average primate data line, is about forty years or so. But our real life span is more than double that.[118] For most of human history, the life expectancy has been much lower, even less than forty years. Until recently, that was due to diseases, wars, and other external causes of death. But some humans always lived to a ripe old age, about the same as today but in fewer numbers. We break all the rules—as primates we break the rules for both mammals and other primates.

Some of the boost for primates, as compared to the mammal average, probably comes from their arboreal habits. In trees, primates are out of the way of hyenas, wild dogs, and lions. Primates in trees can jump between limbs in ways that leopards, for example, cannot. Although primates can be eaten by other primates, such as monkeys by chimps, the issue is not the absolute lack of predators but just the relative amount of freedom from them. We have seen that arboreal lives lead to longer life spans. For example, the average arboreal marsupial mammal lives nearly sixty percent longer than those marsupials stuck on the dangerous ground.

The rest of the boost would likely come from the smarts in the primate brain. Graphing data for primates in terms of life span as a function of brain weight, instead of body weight, in fact gives a tighter correlation. The capacity for evasive behavior in creatures with large brains provides one route to escape from predation. Also, large brains facilitate complex social structures and communication signals useful as warnings. If brain weight is taken to be the crucial factor, humans have it as the top primate. And they possess the greatest longevity. Could those facts be related? I think so.

Taking the chimpanzees and gorillas as models, and given their similar life spans with different body weights (the chimp is the nearest living relative), our last common ancestor with the apes probably had a maximum natural life span of about forty years. But now we have eighty years, give or take. During human evolution we used tools. We had protective social structures. We probably used intelligence to review events by means of mimed stories, which aided the learning and recall about danger, and about opportunities for the future.[119] It is likely that the two major pulses in brain size during human evolution over the past two million years were accompanied (or closely followed by) an increase in life span.

Our life spans derive from our place in nature, honed and changed through evolutionary time. For this we should be grateful to the smarts of our wily ancestors, because their activities brought about conditions of better biochemistry for naturally longer lives. We live via the cumulative success of all those hominids, going back in deep time. In contemplating human biological death we should feel lucky about our natural life span and for all those who, over millions of years, made us what we are today.

LITTLE DEATHS, BIG LIVES

Let's consider a tree. With its stout main branches radiating from the thick core of its trunk, we can first observe the site of early human evolution: primates evolved in such branches. So, to start, honor the tree for some of the reasons we live as long as we do.

What is inside the tree? We can run hands along the thick, corrugated bark. Under all this crackled epidermis lie some vital layers of cells, some dead, some alive. First comes a shallow layer called the phloem. Its vertical columns of cells are alive but are on their way to a planned death. Next comes a thin generative layer that contains the actual living cells of the trunk. These cells reproduce to create all the other layers, both outward and inward. Inside the generative layer is the zone called the xylem. It has columns of tubes and its cells are dead.

The xylem's function is to move the mineral-laden water gathered by the roots up to the needles or leaves. Its special, dead cells are called tracheids. Tracheids (or, when grouped into units, tracheid elements) not only provide water and mineral circulation but also support against gravity. Without tracheids there would be no green life on land except some ground-hugging tiny mosses and soil crust with green bacteria and algae. For not only trees contain tracheids, but so do all nonwoody plants. Tracheids are in all stems, branches, and trunks, in grasses, even in flower stalks (usually in their centers), and in the veins of leaves. Thus without the evolutionary invention of tracheids the land would be virtually deserted.

In tracheids we have an ideal example of how nature turns death into life to create organisms from cells. Tracheids are "functional cell corpses."[120] When tracheids are formed from the generative layer, the replicated living cells destined to become tracheids are emptied of their cell contents. These emptied cells, stiffened with extra cell walls, are then linked end to end in long, vertical bundles that become the water transport tubes. These dead parts remain as functional units within the tree or small plant, absolutely vital to its life.

How do certain cells know how to die and take their places as tracheids? Do they perform the trick by themselves, in suicides? Or are they assassinated by other cells? From what has been seen in laboratories, the process is complex. The favorite cells to study for tracheid formation come from the annual flowering plant the zinnia. In experiments zinnia cells can be induced by hormones and other means to develop into tracheid cells. The main events are as follows.[121]

As the transformation begins, inside the surface of its outer wall a cell destined to become a tracheid lays down an additional shell of cellulose plus a glue called lignin. This so-called secondary wall boasts exquisite patterns of rings, spirals, quilted reticulations, or other tracery that is engineered for long-lasting structural stiffening. A few hours after this secondary wall begins forming, the membrane surrounding the central, fluid-filled vacuole inside the cell disintegrates. This event is a critical point in the process of becoming a tracheid. Enzymes that had been collecting in the fluid of the vacuole, which looks somewhat like a water balloon within the cell, are released into the surrounding living matrix of the cell. Released from the former vacuole, the protein-dissolving enzymes attack specifically targeted proteins. Other enzymes disrupt the DNA and RNA.

After these well-studied events, in the real tree or small plant the dying cell is elevated to the fully functional rank of tracheid. The dissolved mush of cell contents is evacuated and recycled into other cells. Now empty, the dead cell with its species-specific secondary wall takes its place in the water-conducting channels of tracheid elements within the overall xylem tissue.

This is the purest form of sacrifice in which the victim, the cell, becomes a permanent, vitally magnificent part of the whole living creature, the plant.

What about the phloem, sandwiched in between the bark and the generative layer? Unlike the tracheids, the phloem cells are alive. But they carry their own death warrants. Most have been denucleated and thus, without DNA and the attendant ability to synthesize new proteins, the phloem cells need help from companions to live (neighbor cells in this case). In the case of trees, the phloem cells age and die, in annual replacement. While alive they perform the critical function of transporting the viscous, sugary, fluid food made by the leaves to the stems, trunks, and roots, which all need photosynthesized food to live and grow.

Upon death the phloem cells become part of the bark, thus taking on a second job in retirement.

In the case of a tree such as a ponderosa pine, the bark is thick and, as the tree ages, is shed in giant, red-brown flakes. These flakes contain ultrathin layers, which can be peeled off and look like jigsaw puzzle pieces. They are the annual layers of former phloem. The flakes split as the tree trunk expands over the years, creating deep crevasses. Looking into them is akin to viewing the Grand Canyon from an airplane, where the annual layers of former phloem are the sedimentary layers of rock in the canyon.

And so we can say that death makes life in two key places in a tree: in the bark made from dead phloem cells, and in the transport xylem that, when aged, turns into the structural wood. Death is useful. It is not sloughed away but embraced. Death is in fact creative, as sacrificed cells become active participants in the larger living whole, thereby helping all the living cells of the tree.

Controlled cell deaths function in other parts of plants. They occur in zones where ripe fruits and dead leaves cleave away from trees, in a manner that preserves the sealed integrity of the tree. And cell deaths are vital to plant development. Flowers, for example, show functional cell deaths. Once the ovum has been fertilized, the flower petals die. They are not needed, so why keep them supplied with food? Anthers, as well, die after the pollen they grow and store is released. As plant parts are used and then killed, some portions of the cells are returned as nutrients to the main body, a form of auto-cannibalism, a lesson in recycling.

In annual plants, the entire plant body dies after reproduction. Sometimes the dying is controlled in such manner that the seeds receive final boosts of nutrition from the senescing tissues. This sacrifice is of the whole for the part. But this part, the seed, contains the whole, because an entire plant can grow from it, an awesome phenomenon.

We turn now from plants to animals. We have seen that plants utilize death as part of the larger encompassing life, a pattern that the botanical world embodies in a variety of specific ways. This pattern is functional. Is it also found in animals?

Yes, cell death is crucial for all sizes of animals at many stages of their lives. First, consider a number: Every single second in the adult human body, one hundred thousand cells die.[122] This also means that approximately one hundred thousand cells are newborn each second. From these numbers we deduce that on average the body's cells turn over about once a year. These turnover numbers show that life is not just any old matter. It is a whirlpool, a renewing pattern of matter. The atoms can completely change over, but our bodies as systems carry on.

The rates of this turnover vary crucially for different types of cells in our bodies. For example, the brain's cells turn over hardly at all.[123] This stability is probably crucial for the feeling of continuity we have as individuals, so intricately connected are the cells to each other in vast nerve networks. When nerves die, their connections to other nerves in the network vanish, with no guarantee that replacement cells can make the same complex connections. Perhaps memories would shift were brain cells often replaced with ones that made a different pattern of connections!

In contrast, other places in our bodies are hot spots for rapid cell deaths and births. Our skin is one. Its renewal is most apparent in the healing of cuts and scrapes. But the renewal goes on continuously. Everywhere in the skin, below the outermost, scaly layer, new cells are being born. Note the similarity in the way the living cell layer of a tree trunk gives rise to its phloem and bark. The likeness goes deeper. During the development of the outermost skin cells from the underlying living layer, the cells are denucleated. As in the tree, this is a death warrant.

The dying skin cells grow a coat of a specially tough protein called keratin, and lock into geometric positions alongside their neighbors with elaborately fitted borders like dovetail joints. During all this development, these cells are being pushed outward toward the body's surface, where they eventually become the skin we see and feel. There they undergo the ravages of the environment. In sacrifice, they are eventually marched right off the body in continuous shedding. But before being shed they serve as an exquisite protective envelope. A layer of functional cell corpses coats our body.

The gut lining is another hotbed of frenetic cell renewal, with a turnover rate of less than a week. It's a dangerous place for a cell. Food is being digested with a bath of powerful, molecule-ripping enzymes. And there are worlds of bacteria, some of which could be pathogenic. The gut cells are living right on the edge, using chemicals to tear apart things that are not unlike themselves. The place reminds me, as a social analogy, of a coal mine. Workers in the extractive industries often operate at the edge of habitability. They work in dangerous conditions to bring raw materials into society. They contract special diseases and are injured, sometimes fatally. Laws have been increasingly more protective of the worker's rights, but in the early days (think of a Babylonian copper mine) workers were, well, expendable sacrifices like the gut cells.

Then there are our beautiful, little, flexible red blood cells (somewhat like doughnuts, with middle "hole" zones just thinned, not completely punched through). They survive for only about four months after birth, and, like skin cells, they lose their nuclei because they are not going to reproduce.[124] Their lack of nuclei curtails their protein-making capabilities, and consequently their lives, as precisely as does the lack of mouth parts in mayflies. When the red blood cells die of old age, as they lose functions that they cannot maintain without nuclear DNA, they are ingested by phagocytes (various white blood cells). Thus we recycle parts of ourselves.

The immune system also operates via a continuous stream of cell deaths and births. The ravages from serving as internal defense for the body require this. Those cells of the immune system in our blood that ingest and kill viruses and bacteria—in general, the bad guys—have only limited lifetimes themselves. The immune cells are utilized and then they die. But they also die when they go unused for too long.

The outer cell membranes of some of the immune cells are coated with specific types of molecules that can recognize certain patterns of enemies. Hordes of these cells fill ranks for some specific fight, say, against the flu. Other types of immune cells are born with their recognition molecules varied, which provides a way to search for bad guys whose features are not yet known but, in a sense, are guessed at by the immune system. Many cells from this guesswork approach prove useless and the cells die, never recognizing an enemy. Other cells do find enemies, and these hits trigger the production of more of those specific cell types. This organized turnover of the cells provides "improved functioning over time" of the immune system.[125]

In addition, some immune cells born from the speculative approach are potentially dangerous, because they code for attacks on the body itself, like a missile suddenly turning back on its launch site. These, too, are normally eliminated; but autoimmune diseases can result when the deaths of the vagrant cells are not carried out swiftly or neatly enough.

Cells in the body have different life spans, which reflect their lifestyles. As noted, nerve cells for the most part are long-lived. The muscle cells of the heart are also Methuselahs. Pathologist Edmund LeGrand suggests that the reason for this is similar to that for the longevity of nerve cells.[126] The critical timing of the heart, which requires its myocardial cells to conduct electrical impulses and contract, and all within a highly coordinated system, provides little tolerance for replacements that could be ill-fitted for the network. It would be like suddenly shoving me into replacing guitarist Keith Richards of the Rolling Stones, with thirty seconds to go until showtime at Madison Square Garden. Better that Richards himself has a long life and continues to rock and roll.

The cells of the heart and brain need to be protected, because both kinds operate in highly coordinated networks with other cells of their respective organs. But the skin cell on my hand does not directly relate to the skin cell on my cheek. The red blood cell now traveling in my right shoulder doesn't relate to the red blood cell in my little left toe, at least not in the same sense that the interconnected heart cells and brain cells coordinate with others of their kind. The greater or lesser requirement for protecting the integrity of every detail within a subsystem in the body seems to lead to the need for cells with longer or shorter lives. Brain cells are more like the ant queens; they escape predation by being sheltered, so give them potentially long lives. The red blood cells are something like the insects such as fruit flies; they will be damaged no matter what they do, therefore give them short lives.

So far these cell deaths in the human body, as a representative "higher" animal, have been considered as part of routine, day-to-day turnover. But cell death is also key during development into adulthood.

In what is probably the most celebrated example of controlled cell death during human development, the hand is changed from a flat paddle to separate digits, by the deaths of the cells in between zones that will become the future fingers. The general idea here is that cell death is a sculptor of development. Other, general, sculptural processes that can carve with such chisels during animal development include hollowing out solids, pinching off tubes, and fusing sheets.[127] Unlike a sculpture of marble or wood, in which the deleted scraps lie around the floor to be swept up and thrown away, in these cell deaths the material is recycled into other cells.

Cell death is utilized within during development to control not only gross geometry but also the numbers of cells in tissues. Cells are typically overproduced and then culled down to the necessary final numbers. The most famous example of this occurs in the development of our brains. We originally grow about twice as many fetal neurons as eventually survive in the baby's functioning brain. When original neurons develop their connective networks, they compete to reach particular target cells. During the process, these target cells secrete survival chemicals for the neurons groping toward the targets. Those neurons that reach the targets live. They have received enough survival juice. Those that fail to arrive die. This formula serves as a quality control, both on the final number of neurons and as a way to eliminate those who aren't well-connected.

They are fired from the job. This kind of elimination probably works in other tissues as well, to "ensure that cells survive when and where they are needed."[128]

As we saw with the developing neurons, signals between cells are the key to cell death. Some signals are death commands. The classic example occurs in turning a tadpole into a frog. During metamorphosis, the tadpole's thyroid gland secretes a hormone that induces the cells of the tail to die. The tail is eventually reabsorbed and so recycled into the body of what becomes the four-legged, tail-less little frog.

In the Drosophila fruit fly, certain genes have been implicated in the control of cell death during development. Lab experiments demonstrate that these genes—with names such as "grim" and "reaper"—are activated and accordingly make their special protein products an hour or two before certain cell deaths occur in the developing fly. These genes are switched on during development to direct deaths of cells in the sculpting of the larval fly.

In the developing embryo of a chicken, bone-shaping proteins have been identified that deftly eliminate cells between what will become the bird's digits.[129] Experiments with deleting genes in mice have shown that without certain death-dealing genes, the developing mouse dies with a host of problems, including a vast excess of cells in the nervous system.

It is of utmost significance that in these controlled cell deaths all various, internal death signals—from flies to humans—are true signals and not chemical guillotines that murder designated cells. Instead, the remarkable finding is that the guillotines are actually inside the cells themselves. The cells just need to be "told" to die. The protein signals are the messengers. This will become clearer after we also note a second cause of cell death.

Cells can also die if they fail to receive signals telling them to live. How bizarre! The biological finding that cells need support signals from other cells—signals that say, "Yes, stay alive, don't die"—has been a key discovery about cell death, resulting in an avalanche of new research. The awe that the finding has created in the researchers is almost palpable in their papers.

Here are some results of experiments. The neurons whose numbers are trimmed during brain development are an example. To live, a neuron needs signals from the cells that they are contacting (the neurotrophic factor secreted by the target cells that they enervate). In other investigations of "don't die" signals, many cells of animals, particularly higher animals, will die in lab cultures if they are not together in dense enough populations. Death-suppressing signals are made by cells and passed around via diffusion in the culture medium. When these signals drop too low in intensity, the cells die. Low density might be the cell equivalent of loneliness, leading to suicide. In the nematode lab worm, a specific signal that serves to say to cells "You're OK" is a protein called CED-9. If the gene that codes for CED-9 is inactivated by a mutation, most of the worm's cells die and the worm itself dies early in development.

Furthermore, note that it is not just the cells in between future fingers, say, that contain the self-guillotine program. The current theory is that virtually all cells, in many animals at least, contain the self-guillotine. This guillotine is a chemical program that is run to destroy the cell from within by a series of chemical reactions inside the cell. The operation of this program is what has become known as programmed cell death.[130]

What happens to a typical animal cell during programmed cell death is highly organized. Chemicals within the dying cell break the DNA into small packages, which creates a "laddering." Enzymes dissolve other internal parts of the cell, enzymes that the cell itself has created. Some of these enzymes are present all the time. Other types are made when genes are turned on by other molecules in the chemical program. What is called a cascade of molecular events kills the cell. The end result is that the cell separates into a number of smaller, membrane-bound bodies. These bite-sized units are just right for other cells, either neighbors or special cleanup cells, to ingest and hence recycle. Cannibalism of parts for the sake of the whole!

Recycling the material is one function of clean and controlled deaths of healthy cells, such as those between the developing fingers and toes of the vertebrate embryo. Another function is to properly dispose of diseased cells that commit suicide. Programmed cell death is a "clean way of disposing of a cell, one that does not lead to inflammation or spillage of virions."[131] In fact, the process of cell suicide is so neatly discharged that the true, diverse extent of it was missed under the microscopes for many years.

A note on cancer is apropos here. As is well known, cancer cells wildly reproduce out of control. They are the physiological equivalent of someone with a huge ego trying to spread himself or herself far out into the extended social self, to the detriment of the self-realization of others. From the body's point of view, a cancer is a mass of renegade cells, which are deranged, damaged, mutated. It now appears that in cancer cells something has gone wrong not just with their replication programs, but with their death programs. Normally, if a cell suffers a mutation that affects its replicative cycle, then its death program will kick in and lead the cell gently into that good night. Thus the potentially aberrant growth is stopped before it has even started. But if a mutation in the replication program is coupled with a second mutation that cripples the death program, then wild replication can proceed.

Cancer cells do not altogether lack a death program.[132] Many anti-cancer drugs work, at least in part, by activating it. The latent death program is notable, because one would think it beneficial for the cancerous cells to do away with their death programs completely. The cells could theoretically do this, because cancer cells can evolve, becoming ever more virulent in the body by a process of further mutation and selection. The fact that during this evolution to greater malignancy they retain the core of a death program implies that at least portions of the death program must be indispensable to other functions in the cells. More will come in a bit on this topic.

One remarkable finding in aging research over the last couple of decades also has something to do with programmed cell death. The fact is that most, and perhaps all, animals age more slowly if they eat fewer calories. This caloric restriction does not mean that nutrients are lacking from their diet. To the contrary, to work its anti-aging effects the diet must have adequate basic nutrients. But the overall calories are reduced.

How this caloric restriction promotes longevity, better resistance to disease, and overall youthfulness of bodily systems, compared to animals of the same age who have been raised eating as much as they want, is still unknown. One finding is that cells in aging creatures ordinarily lose some of their ability to undergo cell suicide. This is not necessarily good, for as we have seen, cell suicide is necessary for health. Liver cells in rats who have lived under caloric restriction have higher suicide rates. Such rates might help prevent cancers. In fact, it has been suggested that the higher cancer rates as people age might be related to the general loss of their cells to properly run their suicide programs. [133]

Understanding how programmed cell death works will help us fight other diseases as well. Some diseases are exacerbated by overuse of the suicide program. Here we someday might be able to include Alzheimer's and Parkinson's, in which there is aberrant cell death; some researchers suspect the suicide program is faulty. In heart attacks and brain strokes, some cells are killed outright by the massive failures. These cells did not have a chance to run their suicide programs. Cells, however, that are merely close to but not at the centers of damage also die but probably would not have died except for the fact that they ran their suicide programs in response to their milder dose of damage. The hope is for drugs, which, if administered quickly enough, would block the suicide programs of these only slightly damaged cells and thus help general recovery.

Information on aspects of cell suicide in the human body can come directly from research on other creatures, because parallels exist in how their respective cellular death programs work.

In practical terms, say a protein-cutting enzyme is found to be important in the death program of the soil nematode *C. elegans*. Then the database of human enzymes can be searched for similar molecules, judged by similarity of sequences in the amino acids, which are the component parts of enzymes. Some structurally similar human enzyme might be known to exist, but not by function. This would lead scientists to investigate that enzyme more intensively. Perhaps it is a participant in the human cell's chemical cascade of signals or executioner molecules in the cell suicide mechanism.

What parts of the mechanism have been proven alike in very different creatures? For example, in the nematode lab worm, the protein CED-9 is coded by a gene called, appropriately, *ced-9*. As noted, this protein has been shown to inhibit the death program of the worm's cells. Humans possess a gene called *bcl-2*, similar to the worm's *ced-9* gene. The human gene was first known because of the excess production of its Bcl-2 protein in a form of lymphocyte cancer. The excess protein inhibits cell suicide, which helps the cancer, consequently hurting the person.

Now, a worm without the *ced-9* gene lacks the inhibitor CED-9; it will have massive cell suicides throughout its development. It will die. But, if the human *bcl-2* gene is put into the genome of the *ced-9*-deficient worm, the worm will live. The human gene's protein product, Bcl-2, can act as an inhibitor of cell suicide in the worm, in place of the worm's own natural inhibitor.[134]

In this manner aspects of the cell suicide programs, at least in animal cells, have been highly conserved throughout the course of evolution, an awesome physical message. How universal can we go? Can we trace the genetic and operational kinship of cell suicide to build a bridge between animals and plants back in the depths of time?

There are some clear differences between cell suicide in the two kingdoms. One involves the large central vacuole in the plant cells. As noted for the formation of tracheids, this vacuole can accumulate digestive enzymes during the progression of events in cells destined to become tracheids. Disruption of the membrane that separates this vacuole from the active contents of the cell is the point of no return. The released digestive enzymes kill the cell. The collapse of the central vacuole is used not only in the formation of tracheids, but in other places as well.[135] Thus the collapse may be a widespread event during programmed cell death in plants. Because animal cells do not possess the large vacuoles as fundamental cell features, the mechanics of cell suicide between plants and animals must be substantially different.

The facts about similarities and differences in cell suicides between plants and animals will ultimately be proven not by comparing large-scale cell events but via analysis of the molecules of the death machinery, as has been done for CED-9 and Bcl-2. Similar molecules might indicate that the common ancestor of plants and animals—a single-celled creature that lived one or two billion years ago—had components of what went into and remained in the suicide programs for both plants and animals as they evolved. This ancestor might have been able to use special molecules for actual suicides. Alternatively, these molecules might have had other functions in the ancestor cell's life and were found useful for suicides later when plants and animals evolved. Recently, a number of types of molecules important in the programmed cell death of animals have been found in plants, with presumed roles in plant cell suicides. So according to some researchers, parts of a generic suicide program used at least by some animals and some plants was in place in the "crown group" of organisms ancestral to the evolutionary divergence of plants and animals.[136]

The elucidation of specific similarities and differences is on the cutting edge of molecular research, yet already both give us profound understanding. From the similarities we gain insights into how far back in evolution the cell suicide molecules were present. Their existence will help us think about the possible role of controlled cell death long ago in the evolutionary origin of plants and animals. Perhaps such cell death was essential, perhaps a prerequisite, for the evolution of complex body plans of plants and animals.

The differences are at least as, if not more, profound. The differences speak to how the usefulness of death in the service of larger life has been reached through multiple chemical programs in different organisms. They speak to the fact that controlled cell death was a pattern that has been called into being by the process of natural selection in order to fill some function. It is similar to the way that streamlining was discovered by both dolphins and fish as a form that was functional for rapid swimming. Controlled cell death is a process, not a simple shape such as the streamlined body. But chemical programs can be functional just like body shape.

If the need for a function is there and the possibility exists for the molecules to be synthesized and organized to carry out that function, then evolution has at least a fair chance of converting possibility into actuality. Both plants and animals in fact use death to control the development of a large, multicellular system. Parallels in this invented process of functional cell death are somewhat like the way that a style of senescence such as catastrophic senescence was invented in parallel by both plants and animals.

Functional cell death varies in how it is specifically actualized even in a single creature. As an example, consider the leaves of deciduous trees, which crisply fall away before winter. These leaves are not just dropped. To a large extent their cellular contents are recycled back into the plant before their fall. Much of the tough cellulose structure stays in the fallen leaf, of course, but as many nutrients as possible are drawn back into the plant. These nutrients go down into the tree's roots for wintertime storage, which will provide nutrients at next spring's burst of bloom that unfurls and grows the first leaves of the new canopy.

In a study of leaves of five different species of tree, it was found that as they senesce and yellow, the cells of the leaves undergo programmed cell suicide with some key similarities to the classic type of suicide of animal cells.[137] This classic type has, as hallmarks, the condensation and fragmentation of the DNA. Both these phenomena have been confirmed for the cells in the yellowing leaves. This might be parallel or convergent evolution. After all, the DNA probably should be chopped up somehow to make recycling possible. The process, however, has not yet been seen to occur in the formation of tracheids. So a variety of ways apparently exist for the accomplishment of recycling during cell suicide in plants, possibly with some ways maintaining the links to the ancestral cell shared with animals.

There are many forms of cell suicide, yet the greater, overarching idea is the same: Some die for the well-being of the greater whole. Cell death used in the development and daily maintenance of large organisms did not just happen. It was invented during the course of evolution and maintained in so many organisms because it was functional—for example, in creating cell corpses for the transport of water in plants and deleting tissues between the developing digits of vertebrates. Therefore, the issue for evolution became not whether, but how, to effectuate the mechanics of cell death. The "how" was a perfect question for the evolutionary process, which tries out mechanisms and incorporates ones that work into pathways in the molecular dynamics of cells. In functional cell death we find nature clearly speaking a message of "death, thus life."

LIFE AND DEATH
AT THE
SMALLEST SCALE

Given the prevalence and importance of programmed cell death in plants and animals, you probably won't be shocked out of your gourd or skull to discover that programmed cell death also plays a role in the development of another of life's kingdoms: the fungi. Except for the single-celled yeast, upon whose genetics science lavishes attention as a model system (like the nematode and fruit fly), fungi have not been extensively studied for examples of cell suicide. But that is changing.

In species of fungi that produce large mushrooms, these "fruits" often exhibit catastrophic senescence. The mushrooms form out of the rapid growth and clustering of underground white threads (hyphae), which are the distributed "bodies" of fungi. The resulting mushrooms produce spores for the next generations, then die. In fact, the mushroom-making fungi are somewhat like perennial flowering plants. As perennials have roots that remain alive over many seasons underground, the diffuse fungal network can stay alive to send up future rounds of aboveground mushrooms.

A Dutch team of scientists at the Mushroom Experimental Station have studied the formation of the white button mushroom common in supermarkets. The team found that cell death plays a key role in the initial conversion of the subterranean hyphae into a unique formation that kicks off development toward the mushroom.[138]

Some of the hyphae threads become organized within an unusual matrix of material. The hyphae continue growing alongside one other in the matrix and presumably communicate with hormonal signals that alter the expression of their genes toward the goal of that big mushroom in the sky. The new discovery relevant to cell death is that the matrix contains bluish globules loaded with dead fungal cells. Thus some fungal cells died to form this crucial matrix material. The term used earlier for plant tracheids is relevant here for the fungal matrix: functional cell corpses.

These cell deaths occur not as a haphazard mess. As in plant and animal cells that die in the service of development, the fungal cell deaths seem to be precisely timed and genetically programmed. We have an opportunity here to cheer the differences once again, as nature invents variations on a theme.

The fungal cells, in their early deaths as parts of the blue bodies of the matrix, do not show the nuclear fragmentation or condensation characteristic in most animal cell suicides. But rupture of the fungal vacuole membrane is similar to that aspect of plant cell death. Across seven species, the mushroom scientists surveyed a number of developmental events in which cell suicide might play a role, many of which involve the formation of slimy cavities of various sizes. These cavities remind me of how some animals eliminate cells between embryonic fingers and toes. Although the molecular details might differ between animal and fungal cell suicides, the use of death by both to form zones of emptiness is another possible example of a specific functional pattern in the way death supports life.

Detailed similarities in cell death are ultimately established only by analyzing the molecules used in the various execution programs employed by different species. By comparative molecular biology, it is now claimed that some parts of the complex molecular mechanisms in the programmed cell deaths of animals, plants, and fungi are indeed evolutionarily related.[139] These would be the most conserved pieces of the cell death program. One could then search for these mechanisms in today's single-celled organisms. Presumably, at least some of modern protists still carry the molecular processes that existed in life forms from which the animals, plants, and fungi diverged an eon or so ago.

One fascinating case for ancestral death mechanisms has been made by William Clark, who has been studying the potential origins of death in sex.[140] His idea, in a nutshell, is that certain single-celled organisms, such as paramecia, evolved means to "kill" parts of their genomes after mating, and that these techniques could have played roles in the evolution of developmental cell death, which is so necessary for animals, plants, and fungi.

Slipper-shaped paramecia are relatively large cells, about a million times the volume of a bacterium, for example. They are complex, with cilia that they use to swim and with loads of internal structures—their "organs"— some of which can be seen through a microscope. For example, inside the paramecium are two nuclei with DNA: a large macronucleus and a small micronucleus. Paramecia can reproduce by simple cell division, which duplicates the genome exactly in both "daughter" cells, essentially two clones. But a colony of clones eventually senesces. Reproduction runs down.

The problem is apparently the accumulation of mutations in the large macronucleus, which contains multiple copies of the genes in the micronucleus. These copies are needed to support the complexity of the paramecium, but the copies slowly amass faults. Sex can rejuvenate the genome and thus the replication and growth of the colony. This event occurs by genetic exchange between the micronuclei of two genetically distinct paramecia. Think of the micronucleus's DNA as the original copy of the constitution of a country and the macronucleus's DNA as many copies of that constitution. Sex alters the paramecium's constitution in the original storage site. Thus all the now-obsolete copies in the macronucleus must be destroyed and new ones made. That is what the paramecium can do: destroy, recycle, and rebuild the contents of the macronucleus after sex.

Clark's point is that in fragmenting the macronucleus into pieces, the "death of the macronucleus in paramecia that have just had sex is essentially identical to the destruction of the nuclei in mammalian cells" that are undergoing programmed cell death. What we see in the paramecia, and in other single-celled organisms that possess a macronucleus and likewise disintegrate it after sex, could be the evolutionary beginnings of the cell death procedure later used in larger, multicellular organisms. So the logic is that these single-celled organisms, these protists that you can see under a microscope and can watch wriggle and swim around, had the mechanisms that were necessary for tadpoles, mushrooms, and ponderosa trees—and us—to exist.

Another single-celled creature, *Tetrahymena,* also is large and complex, swims using cilia, and carries a large macronucleus for DNA copies, as well as a small micronucleus for code storage. And, like the paramecia, the tetrahymenas eliminate their old macronuclei after sexual conjugation.[141] But tetrahymena cells can do another trick. They can kill themselves.

Tetrahymena cells do not like to live long in low numbers. They prefer to have neighbors, other cells of tetrahymenas. Experiments with cultures of these tiny blimp-shaped swimmers have revealed the existence of critical densities. At populations below these densities, the cells die within a few hours.[142] The deaths occur even in nutritionally luxurious conditions. This implies that the tetrahymena cells know how many other cells are around, probably via chemical signals each cell releases into the local aqueous environment.

From the experiments that have probed how the cells die, researchers conclude that the tetrahymena cells are actively committing suicide. In our bodies, for example, cell suicide requires the synthesis of new enzymes, or additional amounts of current enzymes, to form the substantial network of executioner signals and molecular guillotines that perform the cell destruction. If the cell's ability to perform enzyme synthesis is blocked with chemicals, suicide is thwarted. This is exactly what happens in the cultures of tetrahymena. The blocking of enzyme synthesis drops the death rates of the cells. Therefore, the deaths require active cell processes.

Why would the seemingly independent, free-living cells submit to suicide anyway? Presumably their death is some sort of adaptation that would help them in nature. The answer is not known. Whatever the reason for the obligate urbanity of tetrahymenas, the suicide of single-celled organisms in the context of social regulation is not limited to them. We turn next to the trypanosomatids, which are protozoan parasites that cause such miseries as human sleeping sickness and Chagas disease. These parasites alternate in a life cycle between insect and human hosts.

A discovery vital to the question of the evolutionary driving force of programmed cell death is that several species of trypanosomatids possess the means for cell suicide. Furthermore, the suicide of these single-celled creatures resembles, in several aspects, what has become the classic paradigm for the suicide of animal cells.[143] As in the case of tetrahymena, experiments indicate that signals between the parasitic trypanosomatids regulate the degree to which they live and proliferate, or commit suicide. In other words, how they live or die is density-dependent. In this case the behavior is to increase the rate of suicide when densities climb too high.

Researchers hypothesize that the ability to commit individuals to death is a way of controlling populations. Uncontrolled growth in the parasite populations could lead to the demise of the host before the parasites could complete that portion of their life cycle. While they inhabit a particular insect or vertebrate host, the parasites are creatures in a finite universe. The ability to regulate their populations with the sacrifice of individual cells for the good of other cells in the group would be highly functional.

The trypanosomatids within a particular host (or region within) are mostly clones. This makes it easier for the suicide mechanism to have evolved as a stable characteristic, because then cheats cannot take advantage of those willing to commit suicide. What is a cheat? Imagine a mutant protozoan. It is like the others but without the willingness to take its life when the stage overcrowds. Subsequently, some of the others, the normals, are the only ones that die off when necessary. So the cheat, never dying, would reproduce at a higher rate. However, if the cheat took over, then the cheat could overpopulate, killing the host before the next phase of the cheat's life cycle, and thereby going locally extinct. Populations in which non-suicidal mutants arise are therefore naturally extinguished, whereas suicide-capable populations thrive.

The key to keeping the sacrifice program resistant to invasion by cheats is the relatively tight closure of the system—the host's body, which houses clonal populations of parasites. Now consider multicellular organisms, such as trees, nematodes, fruit flies, and humans. Their cells are basically clones. Different genes are turned on or off, depending on the physiological location and function of a cell as part of a specific tissue—liver, heart, brain, skin. But the basic DNA package in each cell—of an elephant, say—is identical. In the multicellular organisms, cell deaths can be true sacrifices without conflicts of interest between cells of different genomes. In plants and people the sacrificed cells are yielding before their "extended DNA body" of the genetic pattern. Death for life of the whole works most clearly in genetic systems where the parts are all the same. This is also apparently the case in the occurrence of cell suicide in the trypanosomatids and in the tetrahymena, as regulated by density within the colonies. The evolution of programmed death requires a specific relationship between parts and their whole. The parts must not have genetic conflicts of interest with each other.

The cellular slime mold offers another example of this relationship. In the slime molds, amoeba-like protozoa usually live independently as they crawl around and feed within soil or rotting logs. But when food becomes scarce, the cells crawl toward each other and gather into slugs. These slugs, made of perhaps one hundred thousand cells, creep along as a "multicellular" body, then stop and enter into a phase that creates spores. The slug turns into a vertical rising stalk. The leading tip of the stalk swells into a sphere as the stalk lengthens and thins. The whole unit now looks something like a bulbous water tank on top of a columnar tower.

Crucially for the study of cell death, only the amoeba cells in the sphere atop the stalk go on to make spores. These will be taken by the wind to new soil sites where new colonies of amoebas can grow. But what about the cells in the supportive stalk? Those die. They are sacrifices for the cells that make the spores. The stalk cells give their lives to lift the others above the substrate so the spores of those others are propelled into air currents.[144] Here is yet another example of cell suicide in single-celled creatures that is related to a social situation.

Recently, lab experiments using the amoeba slime mold have revealed that cheats can exist.[145] Researchers took cells from the same locale in nature and found genetic variants. These variants could be mixed in randomly paired populations and the resulting mixed colony induced to go into first the slug phase and then the phase of stalk and spore-forming ball. The key discovery was that some variants were able to preferentially put their cells into the ball and not the stalk. These forceful cells were not willing to die as part of the stalk. So in these experimental chimeras, one genome would produce more spores than the other genome, whose cells primarily went into the supporting stalk. This is a case of mixed genomes with potential conflicts of interest. Do the chimeras occur in nature? That is not known. Presumably not. Researchers suspect there are one or more isolating mechanisms that keep the variants apart, thwarting the possibility of cheating in the suicide game. In general, a population that can undergo selective cell death of some of its cells is most likely to be evolutionarily stable when the cells that die are in some sense helping their same genetic selves.

Parasitologist Marcello Barcinski has written,

> *Programmed cell death can occur in any situation where*
> *living cells display features of a structured organization whose*
> *members have developed patterns of relationships and division of*
> *labor through interactions among themselves and/or with*
> *elements of their environment.*[146]

This fits what we have looked at already: clonal colonies of tetrahymena and trypanosomatids where the organizational property of population density affects the health of the cells; genetically identical cell colonies called trees and elephants with division of labor among cells; the genetically identical, normal colonies of the amoeba slime mold, with a division of labor during reproduction. In these cases death is life for others and for the whole. We next turn to the smallest scale of independent cellular life: bacteria.

The small size of bacteria boggles the mind, as do their numbers. Here are just a few factoids: On the skin of the human armpits and groin dwell about a million bacteria per square centimeter. Fortunately for those at all squeamish about so many little critters happily nestled in the crevices of our epidermal cell corpses, most of our skin surface harbors only several thousand bacteria per square centimeter. In our colon, the numbers balloon to around three hundred billion all told. Just one more census result: Globally—including what is within all the ocean water and the soil, as well as inside plants, termites, birds, and other actors in the biosphere—the number of bacteria is estimated as several hundred billion billion billion.[147]

Bacteria are also fecund. Assume that any given single bacterium were to reproduce unchecked by death. In reproduction, when the bacterium has grown to its full size, it simply divides in two. The two daughter cells then each grow until they reach their full sizes, basically double their "birth" weights. Then they each split in two. Starting with a single bacterium, this geometric increase in bacteria weight, by periodic doublings, would create a mass equal to a typical human in two or three days, if the doubling rate is once an hour. And some bacteria can reproduce every twenty minutes in good conditions.

Bacteria in the wilds of soils and natural waters can be quite a bit slower in their cycle of growth and division. An average global estimate for the bacterial doubling time is on the order of once a year.[148] At that rate it would take about sixty years for a single bacterium to expand into a colony of descendants equal to the mass of a human. So then is there no problem with Earth's resources supporting this slow growth? Well, it's not so slow when the bacterial mass is assumed to continue to double. Allowing only for doubling once a year, a colony of bacteria that started equal to a human body would reach a mass equal to that of the entire Earth in only seventy additional years.

Even if the whole Earth were edible, at that far-fetched point when the bacteria ate the planet, the colony would then die. But most of the planet is not edible, so bacteria must die before this ultimate in resource depletion. The same story holds for local sites of limited resources within the biosphere—bacteria must die, at the very least because they run out of resources. Bacteria can starve. They can die from any number of adverse environmental conditions. They have predators. Protozoa and fungi, for instance, feed on them. And some species even prey on others. But is there any cell suicide in bacteria?

According to the reasoning so far, we would need to see something like a society in bacteria, in which deaths of individuals could somehow promote the death genes by helping the social survival of others with those same genes. This is not the case in the traditional view of bacteria as independent, replicating cells. According to microbiologist James Shapiro, the traditional view is based on the medical model, which provided humans with their first scientific experiences of bacteria.[149] In the medical model, a lab culture of pathogenic bacteria can be highly diluted and then spread out on a plate of growth medium. Where single bacteria cells are deposited, they begin to reproduce and grow into colonies, visible as circular zones on the culture medium. Thus the perception has been that single bacteria do just fine and can give rise to vast populations. There would be no driving reason to evolve a cell suicide program.

But Shapiro also points out that these pathogenic bacteria are not the norm and that considering them as such gives rise to the erroneous perception of autonomous replicating units. In fact, it has been known for years that most of the types of bacteria in soil and in natural waters cannot be cultured in labs, or at least not with our current technology. This difficulty derives from the fact that in nature the bacteria live in consortia of different types. Within the consortia the species are tightly interdependent in terms of signals, chemical products, and waste byproducts. They pass around and share so much that separating them from each other kills them. Bacteria are more socially complex than once believed.

Indeed, microbiologist Kim Lewis has recently written, "It is becoming increasingly apparent that bacteria live and die in complex communities that in many ways resemble a multicellular organism."[150] To the extent this is true, we should look for programmed, functional cell death in bacteria populations in cases where we can find clear examples of division of labor in a community situation.

One case is the bacteria called *Myxococcus*. In terms of lifestyle, myxococcus is the bacteria world's version of the amoeba slime mold. These bacteria can aggregate into closely-packed colonies when conditions dictate that they should form spores. The colonies create a stalk with a ball on top. Sound familiar? (In some species, the bacteria's support stalk more closely resembles a tree trunk with multiple branches and "fruits" rather than a single column topped by a sole "water tank.") Cells in the "fruits" will form spores. Cells in the stalk die. Furthermore, during the formation of the colony that will form the stalk and ball, many cells also die by a process known as autolysis. Autolysis is self-dissolution, and in the case of myxococcus the process is required for the formation of the fruiting bodies. The function of the autolysis is not yet known, but it might be a way of supplying additional nutrients to the cells that will become the spores.[151]

Spore formation that induces the death of a helper cell is also seen in the bacterium called *Bacillus subtilis*.[152] A single cell of bacillus can form a spore. This occurs when nutrients drop below a critical level. Spore formation by itself is not so unusual for bacteria. What makes bacillus worth contemplating from the view of functional cell death is that the formation of the spore begins when an original cell copies its genome and builds a wall that separates it into two cell compartments—two cells, really, one large and one small.[153] The small one is the forespore (on its way to becoming a full-fledged spore). The large one is called the mother cell, because it will nurture the development of the spore.

We have in bacillus one of Barcinski's requirements for the emergence of cell death: division of labor. Furthermore, because both were derived from a cell division, they have the same genes. There is, therefore, no genetic conflict of interest, an ideal situation for the evolution of cell suicide. The suicide in fact exists. In the end the mother cell dies. The forespore becomes a spore.

In detail, key assemblies required by the forespore are manufactured by the mother cell. Given the deficient environment, "she" must cannibalize her current proteins to make the special ones necessary for transport into the forespore. The mother cell also needs to manufacture the dissolving enzymes that will attack the dividing wall she herself raised, after the development of the forespore is complete. These dissolving enzymes in turn destroy the remainder of her own wall, dispersing to the environment what is left of her depleted interior contents. She disintegrates.

This so-called autolysis of the mother cell could function to cleanly separate the now useful mother cell from the matured forespore and ensure the integrity of the forespore's shell for its life alone as a spore, and germination sometime later. The mother's autolysis could also release nutrients to neighboring cells, probably kin with the same genes themselves. Alternatively, the autolysis could merely be a forced death due to the breakdown of the mother cell's metabolism after supporting the forespore's development. The last possibility would be death as a byproduct or consequence, not as a functional plan. But Lewis thinks the death in bacillus is most likely a case of programmed death, because "bacteria do not leave much to chance."[154]

In this manner the mother is sacrificed for her single child. Only one can live. The small, completed spore now waits to be shifted somewhere by air or water. It can bide its time for new environmental conditions suitable for emergence from the spore state into a growing bacillus cell once again. During the time the forespore and mother are together, they are like a multicellular organism, with a large body sacrificing itself to bring forth the future, like an annual plant establishing seed and then dying.

A third and final case of programmed cell death in bacteria shows that the dynamics of death on this smallest of living scales can be even more complex than we have seen in myxococcus or bacillus. The creature is *Streptomycetes* (one species of which is the source of the antibiotic streptomycin). It's a genus of colonial soil bacterium. Actually, for streptomycetes, the term colonial is too weak. Microbiologists describe streptomycetes as truly multicellular.

The bacteria in a colony of streptomycetes grow as a "lawn" of threads called hyphae (same name but no relation to the much larger fungal threads), naturally within soil or experimentally in a nutrient dish especially prepared for them by microbiologists. As can be seen under a microscope, after a couple of days of fresh growth, some of the cells differentiate and become aerial. These aerial hyphae project up from the substrate lawn of hyphae and sport spores from certain parts of their bodies. The bacterial lawn changes from waxy yellow during its early growth on nutritious substrate, to powder-white when the aerial hyphae rise up, and then finally to powder-gray when the aerial hyphae create their tough-walled, survival-capsule spores. The term multicellular applies, because streptomycetes show a division of labor into body cells and reproductive cells, with a presumed and intricate system of chemical communication among all.[155]

In the development of reproductive structures in streptomycetes, two waves of programmed cell death occur. The first takes place when the powder-white aerial hyphae rise collectively as ultrathin fluff. Below these many of the substrate hyphae die. Originally, these deaths were thought to result from the disintegration of the outer walls of each hyphal bacterium and the subsequent uncontrolled release of each cell's contents. Instead, the electron microscope has now revealed that cell walls stay relatively intact during death. The nuclear material within each dying cell changes from a relatively concentrated state to a diffuse, loose network. The cell's entire interior changes from a condition with visible texture and discrete component parts into a material that looks homogeneous—perhaps like pureeing a coarse vegetable stew into a uniform blend. The deaths of the cells in the substrate are thus highly controlled, apparently following a set program.

The second round of death occurs in the aerial hyphae as they differentiate into fruiting bodies and move into the powder-gray phase. The tubes of still-living aerial hyphae segment at a series of points and so look something like links of sausages. The spores differentiate within these sausages. The remainder of the hyphae that have not becomes sausages die. The interiors of these sections become blended, a change visible under the microscope. Finally, their contents empty. Some of the hyphae stay fairly well intact and terminate this process as empty, hard-walled tubes. Others shrink and collapse like scrunched, twisted plastic straws.

These deaths are controlled as functional parts of the developmental agenda of the organism. Hyphal demise is orchestrated. Furthermore, the dead cells seem to be functional sacrifices for the living ones. In the first wave, the death of substrate hyphae helps to support the aerial hyphae, and there is likely some transport of the blended innards of the dead cells to the living aerial hyphae on their way to the next stage of development. In the second wave of death, some aerial hyphae die to nurture the spores.

In the cell death of the hyphae of streptomycetes we see similarities to plants. The dead hyphae as functional tubes and even structural supports are like the plant cells that transform into tracheids for structural support, as in the trunk of a tree, and for fluid transport to other cells. In strepto-mycetes there is also a parallel to the amoeba slime molds, where the lower ones in the aggregated colony die to become the stalk that supports the spore ball. A common pattern of death in which some die to support and supply others is manifested in different realms: in bacterial streptomycetes, in the amoeba slime molds (whose crawling cells are a million times a bac-terium's volume), and in plants and the tallest redwood trees. Thus a like pattern of helper cells dying for others has been created by the process of evolution on a number of different scales.

———————

A final role for bacteria death is found not in bacteria themselves, but rather in the ancestors of bacteria still alive today within all cells of our bodies. Here we must include all cells in the bodies of animals, plants, and fungi. We also include the cells of paramecia, tetrahymena, and trypanosomatids. All these examples are the so-called eukaryotic cells, whether free-living or parts of complex organisms. Eukaryotic cells are relative giants compared to bacteria, have a defined nucleus, and are internally complex with organelles. One vital organelle in all eukaryotic cells, without which life as we know it could not exist, is the mitochondrion.

The mitochondrion is descended from a merger—a symbiosis—that occurred something like two billion years ago between a type of bac-terium that was to evolve into today's mitochondrion and a larger, host bacterium. The mitochondrion still carries a small, remnant genome, but most of the original genes have long been transferred into the nucleus of the eukaryotic cell. The symbiosis is indissoluble and essential to the lives of complex cells.

Exactly what the symbiotic proto-mitochondrion and its larger host gave each other in the initial, ancient stages of the symbiosis is not definitively known. But today in the eukaryotic cell we know the functional role of the mitochondrion. It's the energy factory or power plant. Numerous copies of it within each cell oxidize carbon compounds into carbon dioxide and water and create high-energy molecules that can be shunted around inside the cell and used when needed in such processes as protein synthesis and ion transport.

Recently the mitochondrion has given the specialists who study cell death a surprise. They discovered that the mitochondrion is integral to many cases of cell suicide. So far most of the details of what might seem to be this dark aspect of a cell's energy organs have been learned from studies of vertebrates. But the field of research is rapidly expanding. For instance, some suggest mitochondria are involved in the process of programmed cell death not only in many animals but also in plants.[156]

Specifically, a substance called cytochrome-c is leaked from the membrane-bounded mitochondrion at what seems to be the start of the suicide program for many cells committing themselves to death for whatever reason—to form gaps between human fingers, to dissolve a tadpole's tail, to eliminate used or useless immune system cells. This is of great interest because cytochrome-c has been long established as a type of molecule inside the mitochondrion that facilitates the flow of electrons and thereby functions in the mitochondrion's production of energy molecules that are exported into the cell for daily metabolism. But this Dr. Jekyll side has a Mr. Hyde. When cytochrome-c is released from the many mitochondria inside a cell, the cell suicide program goes into action. The events initiated by the mitochondrion might be so crucial that this cell organelle could turn out to be the holy grail of cell suicide, what French scientist Guido Kramer[157] calls the "central executioner."

In addition to cytochrome-c, a substance called Diablo (or, alternatively, Smac) is also released by mitochondria at the start of many cell suicides. Cytochrome-c and Diablo are both promoters of cell suicide, triggering a cascade of other molecular events that eventually dismantle the interior of the cell and cause death. Because the suicides take place so quickly, it has not yet been possible to determine whether or not these chemical releases by mitochondria are the absolute first steps. And in the suicides during the development of the nematode worm, mitochondria are not involved. Neither do mitochondria play roles in insect cell suicides, as studied so far. But the ancestors of these creatures could have dropped a still more ancient role that mitochondria had in the early evolution of cell suicide, as these ancestors evolved other suicide mechanisms. Evidence continues to mount, however, for the central role of the mitochondrion in the generic process of cell suicide.[158]

The findings that support the case of the mitochondrion's importance in cell suicide are abundant enough to lead a number of researchers to begin rethinking the original symbiosis two billion years ago.[159] Perhaps it was not a lovey-dovey bonding of mutual admiration and exchange of necessary chemical gifts. Instead, the proto-mitochondria, as free-living bacteria way back in time, could have forced their way into the large host cells as parasites. Yes, they needed food from the host cells, and yes, perhaps they in turn detoxified the oxygen in the environment for the host cells and eventually provided energy from this detox procedure. But who says the host cells were welcoming? What if the host cells wanted out of the deal? What if the hosts attacked the proto-mitochondria with anti-bacterial toxins, for example?

The answer, which ties into the new findings about the role of the mitochondrion in programmed cell death, is that the proto-mitochondria had means of ensuring their continued existence inside their hosts. For one, they could release cytochrome-c into the body of the host cells. It is not clear how this release could have helped the proto-mitochondria, because if the release killed the cells, the proto-mitochondria would die as well. But perhaps the release only perturbed the cells' chemistry. What good would that have done? If the proto-mitochondria were stressed, perhaps the host cells could be induced to move to a different environment that would better satisfy the proto-mitochondria. Another possibility involves the fact that if the proto-mitochondria lost some of their cytochrome-c via release into the surrounding host cells, the proto-mitochondria would have automatically slowed down their energy activity, one function of which removed oxygen from the host cells (oxygen which was then toxic to the presumed anaerobic host cell). With oxygen leaking in from the environment into the host cell, toxic oxygen radicals would build up in the cell. These would cause a higher rate of mutations in the host cells, some of which, by the luck of the draw, could change some cells into a state that was better for and more friendly to the proto-mitochondria.

These are speculations at this point. But the idea is that the release of what today are death substances from the mitochondria originated as a way not to kill their ancestral host cells but perturb them in ways possibly useful for what were the proto-mitochondria. The release was a way of keeping the host cells in line with the needs of what became the mitochondria.

Over time, host cells that tolerated their proto-mitochondria evolved and survived, and the two lived happily ever after. The symbiosis became obligatory. But it all began with threats, and so programmed cell death in eukaryotes might have begun as a survival defense by the proto-mitochondria. They followed the slogan that the best defense is a good threat. They said to their host cells, in molecular-speak, "I'm here to suck on your wastes, because it's better for me to be inside you and have an exclusive on your wastes than outside competing with others for them after you release the wastes to the dispersing environment. I'll even give you some of my own wastes, and detoxify the environment, which will do you good. But be careful—accept my gift! If you don't like me and try to oust me, or if you try and move somewhere else where you can live without my gift of detoxification, then I'll simply and swiftly hurt you." It is sort of like paying for protection from organized crime, when the protection you're paying for is from them.[160]

Perhaps what we witness today in the mitochondria that perform essential roles in cell suicide is an intensification of function of the original threat molecules—from perturbers to executioners. This is one example of how ancient events of life on Earth could still be present in shaping the rich diversity of life all around us. As we have seen, cell suicide is used in the support of life with spectacular creativity across nature. Because all eukaryotic cells have mitochondria,[161] these organelles must have been essential for the evolution of eukaryotes, meaning everything from paramecia to frogs and redwood trees. Biologist Ichizo Kobayashi[162] has said, referring to the eukaryotic cells, "Their spectacular evolution as multicellular organisms might have become possible only through the gift of death from the parasitic bacteria." If the new views about the mitochondrion hold true, we might have to learn to regard every cell of our body as evidence that an ancient truce against a threat became a useful, necessary component in the molecular machinery of cell suicide that shapes our fingers, weeds neurons from our fetal brains, prunes our immune systems, and daily, in countless ways, keeps us alive.

CONCLUSION:
ETERNITY'S SUNRISE

It's now time both for a bit of summation and for me to step back from the mode of reporting and to fully inhabit flesh and blood. The facts of "death, thus life" in the previous chapters are there for all to contemplate. And I expect all will be affected in uniquely valid ways. But what have I done with these facts? Where do I go from here?

In mid-April 1997, a month had passed since my final exposure to carbon monoxide. During those weeks I had been recovering. No more did I feel like a robot that should be returned to the factory because of serious defects in the actuators and sensors of arms and legs. But neither did I know then that I would not heal all the way.

Accepting a dependence on, and even feeling thankful for, a high-tech drug was in the future during that spring day in the New Mexico mountains alongside a nearby creek. I had reached a particularly lovely spot by hiking a trail that wound upstream for an hour, with a daypack full of papers and notes for working on my nearly completed book about Earth.

Nascent buds on trees were greening the streamside canopy that would soon fill out. In the stream, strands of brilliant green algae swayed in the current, feeding on the early spring's abundant nutrients, released into the flow by wintertime decay and weathering in the high country. Rocks, water, leaves, clouds, and the air itself all sparkled in brilliant sun.

My notebook was open and I was preparing to write. Absent-mindedly I put my hand in the water and began caressing a rock coated with slime. How wonderful it felt. Then I noticed that it was more than wonderful. It was miraculous. It became more then miraculous. It was as if I had never before felt water. As if I had never felt anything my entire life. Soon I forgot all about writing. I embarked upon an orgy of touch, a tactile festival with nature. With my fingers I kissed everything around me, lingering, savoring each detail of serrated, dry grass and convoluted scrap of bark. I would pick up, one after another, an ordinarily trivial little bit of nature and in an extraordinary moment make it the entire universe of an extended self that included both me and it.

Until that moment by the stream, I had not realized how terribly much tactile sensation I had gradually, gradually lost during the months of downward spiral with injured nerves. My hands were awakening from the dead.

I always returned to the water, cold and slippery. I loved the way it caressed the heat away from my hands to carry my warmth downstream. I loved that in this play of thermodynamics my being was entering the world. It was enough to be part of the flow of water and of time. I was being reborn by immersing my hands into the stream of time. At such moments, one becomes extended beyond the skin-bound organism, beyond the world of the senses, to the universe itself.

Where was the creek's source? Geographically, in the mountains ten miles upstream. Prior to that, in the snowfall that was melting in the mountains. But what about the creation of the atoms that compose water? Its hydrogen was made about thirteen billion years ago, early after the big bang that initiated our cosmos. Water's other component, oxygen, was forged in a series of fusions that took hydrogen into heavier elements. This fusing occurred inside the death-and-birth caldrons of stellar cores, and such heavy elements so essential to life were distributed by supernova explosions, the death throes of giant stars.

Here is a case in the realm of physics of "death, thus life." In a sense, the hydrogen dies to form oxygen. This life from death is life of another entity, oxygen in this case, which later combines with some of that abundant, still "living," primordial hydrogen, giving birth to water that now nourishes the algae and my life. In looking at how death becomes life we invariably are obliged to shift scales in types of entities, often upward to the larger, encompassing context. From the giant gas clouds formed from the deaths of stars that burst their contents outward, new stars eventually condense. One was our sun with its circling planets, each formed from star ash that gives us life.

Sitting at that bend along the creek, as I let my fingers deliciously glide over the rocks in the water, I also delighted in the thin layer of slime that coated them. Some of the slime is called a microbial film, with populations more dense, at their own scale, than the densest of human cities. A number of these organisms are probably not too dissimilar to those that lived soon after that enigmatic "origin" of life, at least three and a half billion years ago.

How did life originate? No one knows. We do know the origin of biological death. It began with life. But as we have seen, in biology there is much more to death than simple demise. By the diverse frenzy of organisms actively feeding on the living and on the dead, taking in and putting forth wastes, the entire biosphere is amplified: It is more alive by two hundred times. Death is the resource of life.

Furthermore, death powers evolution. Carl Sagan wrote that the "secrets of evolution are death and time—the deaths of enormous numbers of life-forms that were imperfectly adapted to the environment; and time for a long succession of small mutations that were by accident adaptive, time for the slow accumulation of patterns of favorable mutations."[163] His recipe brings death to the fore in the creative process of evolution. No death, no evolution. No evolution, no forging us and all the other marvels of the living biosphere.

In addition, during the process of evolution organisms have created functional designs that incorporate death as an integral part of what is required to be alive. Scale is crucial. Bacteria with social networks can use death as part of their reproductive efforts to secure the future progress of a portion of their members. And for creatures such as you and me, and cottonwood trees along the streamside, and trout in the chilly water, death is so integral to life that without the ability to utilize death on the scale of internal cells there would be no life for the whole creature.

An appreciation for the role of death in my own being can be traced, via science, back to my biological beginnings. As a human embryo, as a fertilized, hollow sphere with rapidly dividing cells, traveling down my mother's fallopian tube, a major part of me was doomed then to be sacrificed for the being that was to become. The hollow sphere, after it implanted itself in mother's womb, became the placenta. These placental cells lived during "my" development, but eventually died at "my" birth. What became me was a clump of cells inside that hollow sphere, which contained the main layers of cell lineages that would differentiate into my body's parts. I thank those placental cells, who died so that I might live.

I trace some of my genetic heritage for longevity to all those generations of hominids before modern humans. Their success over several million years gave the evolution of their metabolisms bottom-line reasons to develop better repair mechanisms. As a result, we probably live about twice as long (in our intrinsic life spans) as did the last common ancestor of chimpanzees and humans. Metabolic longevity came with the survival boost that resulted from increasing brain size. I thank all those smart, distant ancestors for a longer life.

But I am more. Much more. I am culture. I am a multiplicity of worldviews. I am a thinker, feeler, doer.

For me, one important result from these explorations of death will be ongoing reflection on who and what I am, on how multiple forms of myself are integrated, sometimes even confused and conflicted, and, hopefully, eventually harmonized in daily life. It is difficult to see my ripple in the whole stream of nature and humanity, because that requires being outside to observe. But one can learn to better observe the stream through imagination coupled with understanding. And once the view from the outside is seen, the effects will undulate throughout life.

With my cells, some dying and some being born at a hundred thousand per second, I live. Upon my growth when young, which has turned into senescence with maturity, I contemplate. You and I can use the fact of our personal mortality within a universe of many forms of death as a constant call to more awareness.

You and I can be like Buddha, who saw death by the roadside and was motivated to alter his life in the search for awakening. The death we see, however, does not have to be limited to human death. We can know about bacteria and bigger cells, about birds and pine trees. We can ponder archeological spectacles such as pyramids, as well as the humble, ancient Mimbres grave bowl with punched hole. We can realize that we are the bowls that will be punched. But in the meantime, there is some glorious painting to be done. We paint and create ourselves. Furthermore, this painting will not go unappreciated but can be celebrated by other ripples in the stream.

The awakening experience I had at the creek can be sought in the contemplation of death—as vital addition to our understanding of daily life and life as a whole, sweetening the moment. After working on this book I do not claim to have purged all fears of death, of annihilation. I have anguish. After all, I am human. Such terror is instinctual in reaction to the reality of death that we know. But having details about the workings of death on all scales, and about how it weaves into life, has created in me a wider and deeper appreciation. Now I more consciously aim for gratitude, as a way of life, and for experiences that incorporate scales of knowledge about death and life into my daily perceptions.

While writing this book I was again granted new awareness by flowing waters. I was taking a short walk. I had badly hurt my knee, which was going to mean many weeks of debilitation. More hints of senescence! During one hobbling amble, returning the short distance home I had to cross the shallow, broad river. I stopped in the middle. I liked it there.

I am often annoyed at the number of unbidden thoughts that keep popping into my spotlight of consciousness, motivated by various emotions in my unconscious. But sometimes—and I did this in the middle of the river—I can attach my annoyance (also an emotion) to virtually any thoughts that move from the spotlight and spread back into the unconscious, because most thoughts are so pedestrian. As more such thoughts arrive unbidden back into the spotlight, I attach even more deeply felt annoyance. A loop is established, in which my growing annoyance helps make me more and more aware of the thoughts. Soon—as happened then—the thoughts start to weaken, even vanish. At that moment they were replaced with an intense identification with sensory patterns: clouds, blue sky, and especially, in this case, the tumbling, ever-changing constancy of the river.

Suddenly a phrase went through my mind. Like the earlier thoughts, this one came unbidden. But it was much more attuned to the moment and less to my fretful self. The phrase was from William Blake:[164]

He who kisses the joy as it flies, lives in eternity's sun rise.

My senses in harmony with clouds and liquid turbulence, I began to catch the winged joy. It's out there—see it? I cannot fully describe it—yet it was out there and I was catching it. As I caught it, it flew away, and I caught it again and again, until I was only kissing it—all of everything around me—and every moment became an infinity because I lost awareness of any moments other than the present.

I was immensely satisfied to realize—with a reverberation throughout my being—that when I'm gone this river will still be here. The river is an eternity, compared to my temporality. But I also have the "eternity" of this human present, when I can kiss it. This, I decided later, is what I need to seek as help in facing the prospect of death.

Some vultures were soaring overhead. I can think of them as eternal, too. But they are probably not the same vultures that I first saw here ten years ago. The flow of species is more eternal than I am, and so I see that rivers of species will outlast me. As humans, we almost worship change. We think from parents to children to their children, a pattern in which each stage seems unique. We are much more aware of the finiteness of the human personality than of the relative infinity of species and rivers. Yet we can search for eternity not only in species, rivers, and cultures—but also in the moment. This is a human possibility, because our consciousness is not only of the moment, it can also be aware of an ongoing dawning and evanescence of moments. In short, we realize our transience. So celebrate it. This is the quest. Eternal life right now.

How to achieve this? Everyone is different and does it in their own way. For me, one way is to stop sometimes at a river crossing while on a hike. I dip my hand in the water. It is not to test the temperature. It is to touch eternity. And to have a sunrise of gratitude. Such moments—wordless wonders of unity of all I sense and know—have become my answer to the question, What is death?

NOTES

Photographs: All photos by the author, except for the brain, which is of the author.

Chapter 1
Introduction: Death, Thus Life (pp. 9–23)

1. For principles of patterns, see Volk (1995).
2. Michel De Montaigne. Selected Essays, translated by D.M. Frame. Walter J. Black, New York. 1943. His philosophical approach to his urinary stones is in his essay called "On Experience."
3. Funk, Hoover, and The Jesus Seminar (1997, p. 508), from The Gospel of Thomas (63:1-3). This story is judged "pink" by the Jesus Seminar, which means highly likely that Jesus said something like it. A pink judgment was also given to a similar, but in some ways different, story in Luke (12:16-20).
4. See Eliade's classic (1964) and also Halifax (1982).

Chapter 2
The Three-Pound Miracle (pp. 27–39)

5. LeDoux (1996).
6. Adolphs et al. (1995).
7. Baars (1997, p. 79).
8. Rico (1991, p. 62).
9. Martin et al. (1996).
10. Caramazza et al. (2000).
11. Belin et al. (2000).
12. Koechlin et al. (1999).
13. Smith (1999).
14. Keenan et al. (2000).
15. Keenan et al. (2001).
16. Edelman and Tononi (2000, p. 140).
17. Baars (1997, pp. 27–34).

18. Turner and Knapp (1995). Also, Baars (1997), on the reticular formation: "The function of this system is associated with arousal, and its destruction produces permanent coma." See also Kinney et al. (1994).

19. Kinney et al. (1994).

20. Baars (1997, p. 30). Bogen (1997) also emphasizes a crucial role for the thalamic intralaminar nuclei in consciousness.

21. Edelman and Tononi (2000, p. 216).

Chapter 3
We Live in Two Different Worlds (pp. 41–55)

22. Damasio (1998, p. 1881).

23. One friend, a professor of sociology and philosophy, had an out-of-body journey when he was in his early twenties and visiting New York. He traveled in his mental, astral body from his hotel bed to his house in West Virginia. Floating above his mother, he watched her on the couch, reading a magazine with the television on, until she put the magazine down and fell asleep. He floated down close enough to note the article she had been reading. He was enough of a scientist to want to verify his experience. When he returned home, he asked his mother whether she ever read such-and-such an article. She told him she had been, but didn't finish because she fell asleep. He even found the magazine in a stack on the living room coffee table, where he had seen her lay it.

24. Kris Volk Funk's website is www.newhealthvisions.com.

25. Gardner (1983).

26. Mithen (1999/1996).

27. *National Geographic*, September 1999, p. 70.

28. Firth and Dolan (1996, p. 175).

29. Kosslyn (1994) shows that mental models are continuously being used in processing visual information and thus influence what is "seen." How else could we recognize what we do?

30. Blackmore (1992). In the postscript, she explains how the "astral body" theory of OBEs has basically gone nowhere but that the brain theory has made substantial progress.

31. Blackmore (2001).

32. All quotes are from the article about Newcomb in *Scientific American,* October 1998. The quotes are not citing Newcomb himself.

Chapter 4
The Grateful Self (pp. 57–73)

33. Rhys Davids (1887, p. 83; I took the liberty of substituting "composite" for "component" in the original). In my opinion, the key idea here is that the self, or being, or person is a system of parts (forms of self or components of consciousness). All systems decay. So does the self. In five translations that I have located of Buddha's final words, two use the phrase "compound things," two say "component things," and one uses "composite things." The phrase "component things" was from the older books, and seems awkward and perhaps not even logically correct, given modern usage. But the meaning in all translations was the essentially the same. A fascinating book is *Death, Desire, and Goodness: The Conflict of Ultimate Values in Theravada Buddhism,* by Grace G. Burford (Peter Lang: New York, 1991). She shows that in one of the earliest scriptures, the Buddhist theory of values was developed without the rebirth scheme of later scriptures.

34. Funk, Hoover, and The Jesus Seminar (1997, p. 484), from The Gospel of Thomas (20:2-4).

35. Eliade (1964, p. 62).

36. Baars (1997; this quote and the next are from p. 45).

37. Edelman and Tononi (2000, p. 148).

38. Pollack (1999, chapter 2).

39. Baars makes the case that many higher animals have a form of consciousness, because they possess the requisite physiology. What additional event happened in the evolution of the human brain that gives human consciousness its extraordinary characteristics? No one knows, especially since we don't know what consciousness for different animals is actually like.

40. Phone conversation from January 8, 2001. For background on the self-conscious emotions, see Lewis (1995).

41. Trivers (1985, quotes from pp. 388 and 394).

42. Pinker (1997, p. 404).

43. Angier (2001, p. 38).

44. Goodenough (1998, p. 47).

45. Rue (2000, p. 80).

46. Quoted in McLuhan (1994, p. 177).

47. Ibid., p. 240.

Chapter 5
Nobody Just Dies (pp. 77–98)

48. Forrester (1992).

49. Barley (1997, pp. 54–55).

50. Ibid., pp. 153–155.

51. From material on display at Western New Mexico University Museum, Silver City, New Mexico.

52. For mountain symbolism, see Campbell (1974) on the "world mountain"; for the sun symbolism, Kostof (1995, p. 78).

53. Malinowski 1970/1925, (pp. 49–50).

54. Ibid., p. 51.

55. Panourgiá (1995, p. 143).

56. I want to clarify a point. The above discussion applies to those in one's in-group. History is full of corpse-strewn battlefields and high-level assassinations in which the death of the enemy is hurrahed. Heads stuck on poles and the methods of similar ilk that glorify victory in war, the elimination of the guilty, and other cel- ebrations of bloody murder might be called "anti-funerals" (term suggested by Stephen Power). But for social stability within the kin group, tribe, barony, city- state, nation, or whatever, then those on the inside do not cheer, do not let even one of their own just die. Something must be done. The body must be retrieved, even at great cost. If identification is not possible, it might be honored in a "tomb of the unknown soldier." Again, how the dead within one's group are treated is crucial because the living within that group know that they will

eventually die. Respect as institutionalized ritual in the social memory helps alleviate the potential for terror in the individual and helps implant the knowledge that each matters to the others. Thus the larger cultural life and the death of individuals are intimately bound to each other.

57. Information in the exhibits of the new Greek galleries at the Metropolitan Museum of Art, New York.

58. Malinowski (1970/1925, p.50).

59. Revkin (2000).

60. For information about the Ara Pacis, I follow a technical booklet I picked up there, by E. Bianchi (1998).

61. Kostoff (1995).

62. Rosivach (1994).

63. Burkert (1983, pp. 3–7).

64. Personal communication from Joan Connelly.

65. Leviticus (8:18-21). Holy Bible (1989).

66. Burkert (1983, 50).

67. Ibid.

68. 2 Kings(16:3, 17:17). King Josiah ended these practices: 2 Kings(23:10). Holy Bible (1989).

69. Pringle (1999).

70. Connelly (1996).

71. Ibid., p. 57.

72. Ibid., p. 78.

73. Ibid., p. 71.

74. Leeming and Page (1994, pp. 75–77).

75. "How to pick a ram to make the sacrifice supreme," *New York Times,* March 7, 2001.

Chapter 6
Managing Terror (pp. 101–125)

76. See, for example, Solomon, Greenberg, and Pyszczynski (1998).

77. Loy (1996, p. 3) cites Otto Rank as the originator of the term "immortality projects." The social scientists of terror management theory use this term a lot in association with the pychologist Ernst Becker, whose writing—Pulitzer-Prize–winning book *The Denial of Death* (1973), plus see other works—inspired their original impetus to test out the effects of death awareness on people.

78. Greenberg, Solomon, Pyszczynski (1997, pp. 65, 66).

79. Ibid., p. 65.

80. Solomon, Greenberg, and Pyszczynski (1998, p. 20).

81. Ibid., p. 26.

82. For a very readable overview, see Solomon, Greenberg, and Pyszczynski (1998). For a more technical overview, see Greenberg, Solomon and Pyszczynski (1997). The theory has behind it more than 80 studies, as of Solomon, Greenberg, and Pyszczynski (2000).

83. Arndt et al. (1998).

84. Solomon, Greenberg, and Pyszczynski (1998).

85. That subtle reminders of death are all around us is in Solomon, Greenberg, and Pyszczynski (1998).

86. See Goldenberg et al. (2000).

87. Solomon, Greenberg, and Pyszczynski (1998, p. 40–41).

88. Of course, the question arises whether the tolerant ones can remain so because they have the intolerant folks to wag fingers at when they need some bout of worldview defense as an antidote to death awareness.

89. Greenberg, Solomon, Pyszczynski (1997, p. 131).

90. Loy (1996, chapter 5) provides an analysis of fame and other psychological drives related to seeking forms of immortality as a protection against a general "lack" we feel in life, some of of which is related to mortality.

Chapter 7
Death with Interconnected Dignity (pp. 127–147)

91. Volk, T., Performance of tornado wind energy conversion systems, *Journal of Energy*, 6, 348–350, 1982.

92. Pollack (1999, p. 80).

93. The term "ecology of mind" is from the title of the book *Steps to an Ecology of Mind*, by Gregory Bateson (1972).

94. *Hamlet*, act 5, scene 1.

Chapter 8
Sex and Catastrophic Senescence (pp. 151–163)

95. Finch (1990, see p. 49 and the following pages on adult insects who don't eat). Finch (1998) classifies different types of senescence as rapid, gradual, or negligible.

96. Byrne (2000, p. 243).

97. Kirkwood and Rose (2000).

98. Finch and Rose (1995, see note 12 for salmon). More details can be found in Finch (1990).

99. Finch (1990, pgs. 95–97).

100. Interesting experimental evidence for the disposable body theory comes from experiments on fruit flies. They can be bred for longer lives. The potential is there for them to mutate into flies that senesce more slowly. But in these new flies the reproductive rate goes down. Indeed, when put into competition with flies of the wild, the longer-lived mutants are out-competed.

101. The species is *Phyllostachys bambusoides*. Janzen (1976) says the synchronized seeding produces a bonanza of tasty seeds, more than seed predators can consume, thus ensuring survival of the seeds. With such long intervals between blessed events, no population of seed predators can gear up for complete consumption, which they could do were the seed production of the bamboo population not so intermittent.

Chapter 9
Lifestyle and Life Span (pp. 165–179)

102. The tortoise species is *Geochelone gigantea* and the ocean quahog is *Artica islandica* (Finch, 1998.)

103. Finch (1998).

104. Vasek (1980).

105. *Audubon* magazine, May-June 1998, p. 106. The species is *Armillaria bulbosa*.

106. The idea that one contributing factor for the generally longer lives of large creatures is that their size limits their predators was stated in the concluding remarks by Austad and Fischer (1991, p. B52).

107. Finch (1990, p. 219).

108. Austad and Fischer (1991).

109. Finch (1990, p. 213).

110. Holmes and Austad (1995a, 1995b).

111. Holmes and Austad (1994).

112. Clark (1999).

113. Austad and Fischer (1991).

114. Holmes and Austad (1995a, 1995b).

115. Ogburn et al. (1998).

116. Kirkwood and Austad (2000, p. 235) review other examples of how longer-lived creatures have better cell repair mechanisms than shorter-lived ones. For example, a long-lived rat generates lower levels of cell-damaging reactive oxygen species and higher levels of antioxidant enzymes, compared to a shorter-lived rat species.

117. Finch (1990, p. 268; Fig. 5.5).

118. I take the maximum human life span in such comparisons to be about ninety years, plus or minus. This is less than the 122 years of Jean Calment. However, we have to take into consideration the far larger number of humans available for such statistics, compared to the far fewer numbers of chimps and gorillas, or, say,

elephants and zebras, in zoos. Thus ninety years is an average extraordinary human life span, the possible maximum that I assume would be observed were humans kept in zoos in small numbers, like other creatures.

119. Donald (1995) developed the concept of miming as a step in the evolution of human cognition.

Chapter 10
Little Deaths, Big Lives (pp. 181–199)

120. The term is from Groover and Jones (1999).

121. Fukuda (1997).

122. Vaux and Korsmeyer (1999).

123. LeGrand (1997, see p. 142) points out the need for neurons to be especially stable, for "these cells have spent a lifetime together as part of a network . . . an untrained replacement would not fit in with a team that has taken months or years to master the required skills and cooperation needed to succeed." Some recent findings are negating the dogma of no turnover in the brain, with the discovery that neurons in some parts of the hippocampus can regenerate or turn over. Recently, new neurons have been found in parts of the neocortex of adult macaque monkeys, possibly aiding in learning and memory, functions in which newness could be a benefit. (Gould et al., 1999).

124. Vaux and Korsmeyer (1999, p. 246).

125. LeGrand (1997, p. 141).

126. Ibid., p. 142. A fascinating article that uses a "team player analogy" as a way of thinking about a variety of programmed cell deaths in the body.

127. Jacobson, Weil, and Raff (1997). A fine overview of cell death in animal development.

128. Ibid., p. 351.

129. Ibid.

130. The technical term is "apoptosis" for a hugely encompassing, key category of programmed cell death. Sometimes apoptosis and programmed cell death are used synonymously, and sometimes apoptosis is a subset of programmed cell death, so that's one problem with the word.

131. LeGrand (1997, p. 136).

132. Raff (1998, p. 122).

133. Johnson, Sinclair, and Guarente (1999, p. 298).

134. Raff (1998, p. 121).

135. Fukuda (1997, pp. 686–687).

136. Aravind, Dixit, and Koonin (1999); Aravind, Dixit, and Koonin (2001).

137. Yen and Yang (1998).

Chapter 11
Life and Death at the Smallest Scale (pp. 201–221)

138. Umar and Griensven (1998).

139. Aravind, Dixit, and Koonin (2001) in fact make a case for examples of "death" proteins common to the ancestor of fungi, plants, and animals.

140. Clark (1996, 1999).

141. Davis et al. (1992).

142. Christensen et al. (1995).

143. Welburn, Barcinski, and Williams (1997) present an overview of the findings, some of which appear in Welburn et al. (1996) and Ameisen et al. (1995).

144. Cornillon et al. (1994).

145. Kessin (2000) and Strassmann, Zhu, and Queller (2000). The question that remains from the slime mold studies is how the variants that end up mostly in the stalks of the mixed colonies continue to survive in nature. They should be selected against. Presumably in nature there are mechanisms that allow isolation to avoid the competition that would be disastrous for those variants made to go into the sacrificial stalk positions by the more dominant variants.

146. Barcinski (1998, p. 21).

147. Whitman, Coleman, and Wiebe (1998). The count does not include the land below eight meters or the ocean sediments below ten centimeters.

148. Ibid.

149. Shapiro (1998).

150. Lewis (2000, p. 503).

151. Ibid.

152. Losick and Dworkin (1999) and Lewis (2000).

153. Chaloupka and Vinter (1996).

154. Lewis (2000, p. 504).

155. Miguélez, Hardisson, and Manzanal (1999).

156. Jones (2000).

157. Kramer (1997). He specifically calls the event of permeability transition in the membrane of the mitochondrion the "central executioner," or the "death switch."

158. Finkel (2001) covers the science on the status of the central role of mitochondria in programmed cell death, pros and cons.

159. Blackstone and Green (1999), Kobayashi (1998), Kroemer (1997).

160. A related issue is what are called "addiction modules" in some bacteria. They are certain kinds of tiny circular bits of DNA (plasmids), which lie outside the core DNA of the bacterium. The addiction modules code for a slowly decaying toxin and a fast-decaying antidote to the toxin. If the bacterium tries to oust the addiction-module, neither the toxin nor its antidote continue to be made. The antidote decays quickly. This leaves the slower-decaying toxin in the cell, killing the cell. In this way, some cells are killed, but the modules remain in those cells that allow the module to continue undisturbed. In fact, one theory is that programmed cell death in bacteria, in general, began with, and then evolved from, such addiction modules. See Yarmolinsky (1995).

161. Except in special cases.

162. Kobayashi, (1998, p. 373).

Chapter 12
Conclusion: Eternity's Sunrise (pp. 223–230)

163. Sagan (1985, p. 20).

164. "Eternity," by William Blake: "He who binds to himself a joy/Does the winged life destroy/But he who kisses the joy as it flies/lives in eternity's sun rise."

BIBLIOGRAPHY

Adolphs, Ralph; Daniel Tranel; Hanna Damasio; and Antonio Damasio. Fear and the human amygdala. *The Journal of Neuroscience, 15,* 5879-5891, 1995.

Ameisen, Jean Claude, et al. Apoptosis in a unicellular eukaryote *(Trypanosoma cruzi)*: Implications for the evolutionary origin and role of programmed cell death in the control of cell proliferation, differentiation and survival. *Death and Differentiation, 2,* 285-300, 1995.

Angier, Natalie. Confessions of a lonely atheist. *The New York Times Magazine.* January 14, 34-38, 2001.

Aravind, L., V. M. Dixit, and E. V. Koonin. Apoptotic molecular machinery: Vastly increased conplexity in vertebrates revealed by genome comparisons. *Science, 291,* 1279-1284, 2001.

Aravind, L., Vishva M. Dixit, and Eugene V. Koonin. The domains of death: Evolution of the apoptosis machinery. *Trends in Biochemical Sciences, 24,* 47-53, 1999.

Arndt, Jamie; Jeff Greenberg; Linda Simon; Tom Pyszczynski; and Sheldon Solomon. Terror management and self-awareness: Evidence that mortality salience provokes avoidance of the self-focused state. *Personality and Social Psychology Bulletin, 24,* 1216-1227, 1998.

Austad, Steven N., and Kathleen E. Fischer. Mammalian aging, metabolism, and ecology: Evidence from bats and marsupials. *Journal of Gerontology: Biological Sciences, 46,* B47-B53, 1991.

Baars, Bernard J. *In the Theater of Consciousness: The Workplace of the Mind.* Oxford University Press: New York. 1997.

Barcinski, Marcello A. Apoptosis in trypanosomatids: Evolutionary and phylogenetic considerations. *Genetics and Molecular Biology, 21,* 21-24, 1998.

Barley, Nigel. *Grave Matters: A Lively History of Death Around the World.* Henry Holt: New York. 1997.

Bateson, Gregory. *Steps to an Ecology of Mind.* New York: Ballantine. 1972.

Belin, Pascal, et al. Voice-selective areas in human auditory cortex. *Nature, 403,* 309-312, 2000.

Bianchi, Emanuela. *Ara Pacis Augustae.* Rome: Fratelli Palombi. 1998.

Blackmore, Susan. Giving up the ghosts: End of a personal quest. *Skeptical Inquirer,* *25* (March/April), 25, 2001.

Blackmore, Susan J. *Beyond the Body: An Investigation of the Out-of-the-Body Experiences.* Chicago: Academy. 1992 (with new postscript).

Blackstone, Neil W., and Douglas R. Green. The evolution of a mechanism of cell suicide. *BioEssays, 21,* 84-88, 1999.

Bogen, Joseph E. Some neurophysiologic aspects of consciousness. *Seminars in Neurology. 17,* 95-103, 1997.

Burkert, Walter. *Homo Necans: The Anthropology of Ancient Greek Sacrificial Ritual and Myth.* (translated by Peter Bing) University of California Press: Berkeley. 1983/1972(orig.)

Byrne, Richard W. Evolution of primate cognition. *Cognitive Science, 24,* 543-570, 2000.

Calvin, William H. *How Brains Think: Evolving Intelligence, Then and Now.* BasicBooks: New York. 1996.

Campbell, Joseph. *The Mythic Image.* Princeton University Press: Princeton, New Jersey, 1974.

Caramazza, Alfonzo, et al. Separable processing of consonants and vowels. *Nature, 403,* 428-430, 2000.

Chaloupka, J., and V. Vinter. Programmed cell death in bacteria. *Folia Microbiologica, 41,* 451-464, 1996.

Christensen, Søren T., et al. Mechanisms controlling death, survival and proliferation in a model unicellular eukaryote *Tetrahymena thermophila. Death and Differentiation, 2,* 301-308, 1995.

Clark, William R. *A Means to an End: The Biological Basis of Aging and Death.* Oxford: Oxford University Press. 1999.

————. *Sex and the Origins of Death.* Oxford: Oxford University Press. 1996.

Connelly, Joan B. Parthenon and Parthenoi: A mythological interpretation of the Parthenon frieze. *American Journal of Archaeology, 100,* 53-80, 1996.

Cornillon, Sophie, et al. Programmed cell death in *Dictyostelium. Journal of Cell Science, 107,* 2691-2704, 1994.

Damasio, Antonio R. Investigating the biology of consciousness. *Philosophical Transactions of the Royal Society of London. (B) 353*, 1879-1882, 1998.

Davis, Maria C., et al. Programmed nuclear death: Apoptotic-like degradation of specific nuclei in conjugating *Tetrahymena. Developmental Biology, 154*, 419-432, 1992.

Donald, Merlin. *Origins of the Modern Mind: Three Stages in the Evolution of Culture and Cognition.* Harvard University Press: Cambridge, MA. 1991.

Edelman, Gerald M., and Giulio Tononi. *A Universe of Consciousness: How Matter Becomes Imagination.* New York: Basic Books. 2000.

Eliade, Mircea. *Shamanism: Archaic Techniques of Ecstasy.* Princeton University Press: Princeton. (Bollingen Series LXXVI) 1964.

Finch, Caleb E. Variations in senescence and longevity include the possibility of negligible senescence. *Journal of Gerontology, 53A*, B235-239, 1998.

———. *Longevity, Senescence, and the Genome.* The University of Chicago Press: Chicago. 1990.

Finch, Caleb E., and Michael R. Rose. Hormones and the physiological architecture of life history evolution. *The Quarterly Review of Biology, 70*, 1-52, 1995.

Finkel, Elizabeth. The mitochondrion: Is it central to apoptosis? *Science, 292*, 624-626, 2001.

Firth, Chris, and Ray Dolan. The role of the prefrontal cortex in higher cognitive functions. *Cognitive Brain Research, 5*, 175-181, 1996.

Forrester, R. E. *Archeological excavations at the X-S-X Ranch*, northern Grant County, New Mexico. September, 1992.

Fukuda, Hiroo. Programmed cell death during vascular system formation. *Cell Death and Differentiation, 4*, 684-688, 1997.

Funk, Robert W., Roy W. Hoover, and The Jesus Seminar. *The Five Gospels: The Search for the Authentic Words of Jesus.* HarperSanFrancisco. 1997.

Gardner, Howard. *Frames of Mind: The Theory of Multiple Intelligences.* New York: BasicBooks. 1983.

Goldenberg, J. L., et al. Fleeing the body: A terror management perspective on the problem of human corporeality. *Personality and Social Psychology Review, 4*, 200-218, 2000.

Goodenough, Ursula. *The Sacred Depths of Nature.* New York: Oxford U. Press. 1998.

Gould, Elizabeth, et al. Neurogenesis in the neocortex of adult primates. *Science, 286*, 548-552.

Greenberg, Jeff, Sheldon Solomon, and Tom Pyszczynski. Terror management theory of self-esteem and worldviews: empirical assessments and conceptual refinements. *Advances in Experimental Social Psychology*, *29*, 61-139, 1997.

Groover, Andrew, and Alan M. Jones. Tracheary element differentiation uses a novel mechanism coordinating programmed cell death and secondary cell wall synthesis. *Plant Physiology*, *119*, 375-384, 1999.

Halifax, Joan. *Shaman: The Wounded Healer*. Thames and Hudson: New York. 1982.

Holmes, Donna J., and Steven N. Austad. The evolution of avian senescence patterns: Implications for understanding primary aging processes. *American Zoologist*, *35*, 307-317, 1995a.

———. Birds as animal models for the comparative biology of aging: A prospectus. *Journal of Gerontology: Biological Sciences*, *50A*, B59-B66, 1995b.

———. Fly now, die later: Life-history correlates of gliding and flying in mammals. *Journal of Mammology*, *75*, 224-226, 1994.

Holy Bible. World Publishing: Grand Rapids, Michigan. 1989.

Jacobson, Michael D., Miguel Weil, and Martin C. Raff. Programmed cell death in animal development. *Cell*, *88*, 347-354, 1997.

Janzen, Daniel H. Why bamboos wait so long to flower. *Annual Reviews of Ecological Syst.*, *7*, 347-391, 1976.

Johnson, F. Brad, David A. Sinclair, and Leonard Guarente. Molecular biology of aging. *Cell*, *96*, 291-302, 1999.

Jones, Alan. Does the plant mitochondrion integrate cellular stress and regulate programmed cell death? *Trends in Plant Sciences*, *5*, 225-230, 2000.

Keenan, Julian P., et al. Self-recognition and the right prefrontal cortex. *Trends in Cognitive Sciences*, *4*, 338-344, 2000.

———. Self-recognition and the right hemisphere. *Nature*, *409*, 305, 2001.

Kessin, Richard H. Cooperation can be dangerous. *Nature*, *408*, 917-919, 2000.

Kinney, Hannah C., et al. Neuropathological findings in the brain of Karen Ann Quinlin (The role of the thalamus in the persistent vegetative state). *The New England Journal of Medicine*. *330*, 1469-1475, 1994.

Kirkwood, Thomas A., and Steven N. Austad. Why do we age? *Nature*, *408*, 233-238, 2000.

Kobayashi, Ichizo. Selfishness and death: Raison d'etre of restriction, recombination and mitochondria. *Trends in Genetics*, *14*, 368-374, 1998.

Koechlin, Etienne, et al. The role of the anterior prefrontal cortex in human cognition. *Nature*, *399*, 148-151, 1999.

Kosslyn, Stephen. *Image and Brain: The Resolution of the Imagery Debate*. Cambridge, Massachusetts: The M.I.T. Press. 1994.

Kostof, Spiro. *A History of Architecture: Settings and Rituals*. Oxford University Press. Second Edition. 1995.

Kroemer, Guido. Mitochondrial implication in apoptosis. Towards an endosymbiotic hypothesis of apoptosis evolution. *Cell Death and Differentiation, 4*, 443-456, 1997.

LeDoux, Joseph. *The Emotional Brain: The Mysterious Underpinnings of Emotional Life*. Simon & Schuster: New York. 1996.

Leeming, David, and Jake Page. *Goddess: Myths of the Female Divine*. New York: Oxford University Press. 1994.

LeGrand, Edmund K. An adaptationist view of apoptosis. *The Quarterly Review of Biology, 72*, 135-147, 1997.

Lewis, Kim. Programmed death in bacteria. *Microbiology and Molecular Biology Reviews, 64*, 503-514, 2000.

Lewis, Michael. Self-conscious emotions. *American Scientist, 83*, 68-78, 1995.

Losick, Richard, and Jonathan Dworkin. Linking asymmetric division to cell fate: teaching an old microbe new tricks. *Genes and Development, 13*, 377-381, 1999.

Loy, David. *Lack and Transcendence: The Problem of Death and Life in Psychotherapy, Existentialism, and Buddhism*. Humanities Press: Atlantic Highlands, New Jersey. 1996.

Malinowski, Bronislaw. Magic, science, and religion. In *Science, Religion, and Reality* (J. Needham, ed.). Port Washington, NY: Kennikat Press. 1970, re-issue from 1925.

Martin, Alex, et al. Neural correlates of category-specific knowledge. *Nature, 379*, 649-652, 1996.

McLuhan, T. C. *Touch the Earth: Encounters with Nature in Ancient and Contemporary Thought*. New York: Simon & Schuster. 1994.

Miguélez, Elisa M., Carlos Hardisson, and Manuel B. Manzanal. Hyphal death during colony development in *Streptomyces antibioticus*: Morphological evidence for the existence of a process of cell deletion in a multicellular prokaryote. *The Journal of Cell Biology, 145*, 515-525, 1999.

Mithen, Steven. *The Prehistory of the Mind: The Cognitive Origins of Art, Religion, and Science*. London: Thames and Hudson. 1999/1996.

Nehrer, Andrew. *The Psychology of Transcendence*. New York: Dover. 1990/1980.

Ogburn, Charles E., et al. Cultured renal epithelial cells from birds and mice: Enhanced resistance of avian cells to oxidative stress and DNA damage. *Journal of Gerontology: Biological Sciences*, 53A, B287-B292, 1998.

Panourgiá, Neni. *Fragments of Death, Fables of Identity: An Athenian Anthrography*. Madison, WI: University of Wisconsin Press. 1995.

Pinker, Steven. *How the Mind Works*. New York: W.W. Norton. 1997.

Pollack, Robert. *The Missing Moment: How the Unconscious Shapes Modern Science*. New York: Houghton Mifflin. 1999.

Pringle, Heather. Temples of Doom. *Discover*, 78-85, March 1999.

Raff, Martin. Cell suicide for beginners. *Nature*, *396*, 119-122, 1998.

Revkin, Andrew C. U.S. plan would sacrifice baby eagles to Hopi ritual. *New York Times*, Oct. 29, 2000, 1.14.

Rico, Gabriele Lusser. *Pain and Possibility: Writing Your Way Through Personal Crisis*. Los Angeles: Tarcher. 1991.

Rhys Davids, T. W. *Buddhism: Sketch of the life and teachings of Gautama, the Buddha*. London: Society for Promoting Christian Knowledge. 1887.

Rosivach, Vincent J. *The System of Public Sacrifice in Fourth-Century Athens*. Atlanta: Scholar's Press. 1994.

Rue, Loyal. *Everybody's Story: Wising Up to the Epic of Evolution*. Albany, NY: State University of New York Press. 2000.

Sagan, Carl. *Cosmos*. New York: Ballantine. 1985.

Shapiro, James A. Thinking about bacterial populations as multicellular organisms. *Annual Reviews of Microbiology*, *52*, 81-104, 1998.

Smith, Huston. *The World's Religions*. San Francisco: HarperSanFrancisco. 1992.

Solomon, S., J.Greenberg, and T. Pyszczynski. Pride and prejudice: Fear of death and social behavior. *Current Directions in Psychological Science*, *9*, 200-204, 2000.

Solomon, Sheldon, Jeff Greenberg, and Tom Pyszczynski. Tales from the crypt: On the role of death in life. *Zygon Journal of Religion and Science*, *33*, 9-43, 1998.

Strassmann, Joan E., Y. Zhu, and D.C. Queller. Altruism and social cheating in the social amoeba *Dictyosteliuym discoideum*. *Nature*, *408*, 965-967, 2000.

Trivers, Robert. *Social Evolution*. Menlo Park, CA: Benjamin/Cummings. 1985.

Turner, Blair H. and Margaret E. Knapp. Consciousness: a neurobiological approach. *Integrative Physiological and Behavioral Science*, *2*, 151-156, 1995.

Umar, M. Halit, and Leo J. L. D. Van Griensven. The role of morphogenetic cell death in the histogenesis of the mycelial cord of *Agaricus bisporus* and in the development of macrofungi. *Mycological Research*, *102*, 719-735, 1998.

Vasek, Frank C. Creosote bush: Long-lived clones in the Mojave Desert. *American Journal of Botany*, *67*, 246-255, 1980.

Vaux, David L., and Stanley J. Korsmeyer. Cell death in development. *Cell*, *96*, 245-254, 1999.

Volk, Tyler. *Gaia's Body: Toward a Physiology of Earth*. Copernicus Books/Springer-Verlag: New York. 1998.

———. *Metapatterns Across Space, Time, and Mind*. New York: Columbia University Press. 1995.

Welburn, Susan C., M. A. Barcinski, and G. T. Williams. Programmed cell death in trypanosomatids. *Parasitology Today*, *13*, 22-26, 1997.

Welburn, Susan C., et al. Apoptosis in procyclic *Trypanosoma brucei rhodesiense*. *Death and Differentiation*, *3*, 229-236, 1996.

Whitman, William B, David C. Coleman, and William J. Wiebe. Prokaryotes: The unseen majority. *Proceedings of the National Academy of Sciences, U.S.A.*, *95*, 6578-6593, 1998.

Yarmolinsky, Michael B. Programmed cell death in bacterial populations. *Science*, *267*, 836-837, 1995.

Yarrow, H. C. *North American Indian Burial Customs*. (Edited by V. LaMonte Smith) Eagle View: Ogden, Utah. 1988 (originally 1878).

Yen, Cheng-Hung, and Chang-Hsien Yang. Evidence for programmed cell death during leaf senescence in plants. *Plant and Cell Physiology*, *39*, 922-927, 1998.

ACKNOWLEDGMENTS

I was fortunate to have two editors. First, Peter N. Nevraumont helped the project progress through its many stages. Second, though he came into it after the first draft, Stephen S. Power, at John Wiley & Sons, approached the project with the same diligence and care as if it had been his to oversee from the start. I thank both Peter and Stephen for their essential contributions, and acknowledge Susan Bernat for refined copyediting.

I also wish thank all those who aided me in myriad ways, through advice, discussions, or reading text: David Abram, Amelia Amon, Jamie Arndt, Connie Barlow, Ace Barnes, Jean Barnes, Tim Binkley, Susan Blackmore, Jackie Brookner, Allen Campbell, Carla Campbell, Orren Champer, Vivian Champer, William Clark, Joan Breton Connelly, Susan Doll, Harwood Fisher, Martha Foley, Kristin Volk Funk, Mary Gordon, Kathelin Gray, Jeff Greenberg, Ricardo Guerrero, Nancy Hartel, Iris Hoffert, Marty Hoffert, Curtiss Hoffman, Lyn Hughes, Ralph Klicker, Aaron Krochmal, Joseph LeDoux, Michael Lewis, Emily Loose, Ann Marek, Gabby Marek, Lynn Margulis, Andrew Neher, Jill Neimark, Ann Perrini, Robert Pollack, Michael Rampino, John Richards, Susan M. Richards, John Richardson, Gabriele Rico, Bill Rowe, William Ruddick, Claude M. Scales III, Saundra Schimmelpfennig, William Schuster, Sonya Shannon, Sandy Silky, Michael Simpson, Sheldon Solomon, Guenther Stotzky, Francesco Tubiello, Richard Turner, Sylvia Turner, Janice Volk, Jim Volk, Ken Volk, Lauren Volk, Tom Volk, Luke Wallin, Deborah Winiarski.

Finally, I thank the staff at the interlibrary loan department of Miller Library at Western New Mexico University, the information office at the Arizona-Sonoran Desert Museum, the numerous researchers who sent me their papers, and all the scholars across so many fields whose work guided and informed my efforts.

INDEX